DIVIDED NEIGHBORHOODS

Volume 32, URBAN AFFAIRS ANNUAL REVIEWS

DIVIDED NEIGHBORHOODS

Changing Patterns of Racial Segregation

Edited by
GARY A. TOBIN

Volume 32, URBAN AFFAIRS ANNUAL REVIEWS

To Gussie,
whom I love very much.

For information address:

SAGE Publications, Inc.
2111 West Hillcrest Drive
Newbury Park, California 91320

SAGE Publications Ltd.
28 Banner Street
London EC1Y 8QE
England

SAGE Publications India Pvt. Ltd.
M-32 Market
Greater Kailash I
New Delhi 110 048 India

Printed in the United States of America

Library of Congress Cataloging-in-Publication Data

Divided neighborhoods: changing patterns of racial segregation/edited by Gary A. Tobin.
p. cm.—(Urban affairs annual reviews; v. 32)
Bibliography: p.
ISBN 0-8039-2670-7 ISBN 0-8039-2671-5 (pbk).
1. Discrimination in housing—United States. 2. Housing policy—United States. 3. United States—Race relations. 4. Afro—Americans—Housing. I. Tobin, Gary A. II. Series.
HT108.U7 vol. 32
[HD7288.76.U5] 87-20735
363.5′1—dc19 CIP

SECOND PRINTING, 1990

Contents

Preface

I WANT TO EXPRESS my gratitude to my teachers over the years, who helped me think about the things that matter: Ed Greenwood, George Glass, Walter Ehrlich, and Glen Holt. A special nod is reserved for Iris Markman. Many thanks to the people with whom I have worked on these issues for many years: Ernest Calloway, Tim Barry, Jesse Horstman, Lou Berra, David Tatel, Arthur Benson, Teddy Shaw, Jim Liebman, and many of the authors in this volume. Larry Sternberg guided the logistics of this publication, and was helpful as always. Sylvia Riese was terrific in making sure this manuscript reached the publisher. She has learned more about housing segregation than she ever wanted to know. I tip my hat to her once again.

—Gary A. Tobin

Introduction:
Housing Segregation in the 1980s

GARY A. TOBIN

PATTERNS OF HOUSING SEGREGATION in the 1980s differ from those of the 1950s or the 1940s, but they are no less real. More than twenty years have passed since the landmark open housing legislation of the 1960s. Yet the vast majority of the nation's minorities, particularly blacks, remain locked in segregated neighborhoods throughout the United States. Segregation remains, largely as a function of continued discrimination, albeit usually more subtle than the discrimination of the past. Persistent patterns of racial segregation are also the legacy of the actions in both the public and private sectors. Discrimination was institutionalized into housing markets throughout most of this century. Although some discriminatory practices have been largely abandoned, or greatly reduced, others have become integral components of neighborhood selection, sales, and financing. The past restrains the present, but segregative actions also continue unabated.

It would be simplistic, however, to argue that nothing has changed in the past forty years. Racially restrictive covenants are no longer enforced in the courts; Realtors rarely openly refuse to show blacks houses in certain neighborhoods; and federally subsidized housing is no longer sanctioned to be racially segregated. At least on paper, through legislative and some judicial and executive actions, housing discrimination, and the subsequent segregation of neighborhoods, constitutes a violation of civil rights in the 1980s.

But two myths about the nature of housing segregation continue today. Both of them should be finally put to rest. The first myth holds that great strides have been made every decade in reducing the overall segregation of minorities. Especially in the post-1960s era, it is believed that everybody could live where they choose, and therefore "past" problems have been resolved. The second myth holds that where housing segregation has not disappeared, it is a function of two factors. The first is that minorities are poor, and housing markets segmented by income, find racial segregation as an artifact. Second, minorities cluster together because they are showing a revealed preference for living among people of their own kind.

Overwhelming evidence has always refuted these arguments. Poor whites are no more likely to live with poor blacks than upper middle-income blacks are to live with upper middle-income whites. If class or income were really influential in patterns of racial segregation, then the levels of racial isolation should have been reduced a great deal some time ago. Of course, where the income argument fails, the revealed preference argument is introduced. Whites prefer not to live with blacks, even though blacks may prefer to live with some whites, and therefore dislocation in the market is always present. In short, whites simply do not want to live with blacks in great enough proportions to support integrated neighborhoods.

Most whites do indeed avoid black neighbors. But these preference arguments have always failed to take into account the role that institutions played in molding racial attitudes. The history of both governmental and private sector actions, a partnership of discriminatory practice throughout most of the twentieth century, set standards in which individuals made their housing choices. Racially restrictive covenants were enforced by the state, utilized by FHA as a means for evaluating neighborhood stability, and utilized by bankers, insurers, and realtors in everyday housing market operations. Federally subsidized housing was segregated by race. In many states, public schools were legally segregated, reinforcing existing housing patterns and vice versa. Blacks and other minorities were openly and blatantly refused housing in white areas.

By the 1950s, the patterns of racial segregation had been institutionalized to the point that the entry of a black into a white neighborhood signaled a rapid transition that often resulted in panic. By helping to construct a totally segregated housing system, governments at all levels and often housing market actors such as realtors, insurers, and bankers, helped mold straightforward and rigid guidelines for the housing consumer. The system was simply understood: If blacks or other minorities entered a neighborhood, they must be either driven out, which was often the case, or white residents should begin looking for housing elsewhere. Both of these practices continue today. In this system, individual choice becomes a nonsensical notion. Segregation could give way only to resegregation, from white to black. The sooner the white residents left a transitional neighborhood, the less risk they incurred in terms of property values declining, increased crime, and deterioration of the public schools. Purchasing a home in a transitional area also constituted a big risk. For blacks, a great "neighborhood chase" was perpetually under way, looking for stable neighborhoods that were by their very nature destabilized the moment they entered them. The legacy of these practices continues to frame housing market systems in the 1980s.

In his article entitled "Choosing Neighbors and Neighborhoods: The Role of Race in Housing Preference," Joe Darden effectively dismisses the preference for one's own kind argument. He traces the literature, and reveals the flaws in these arguments. In "Housing Market Discrimination and Black Suburbanization in the 1980s," John Kain shows that income is a very poor explanation for patterns of racial segregation. John Yinger ties these two pieces together in "The Racial Dimension of Urban Housing Markets in the 1980s." Together, the three authors definitively demonstrate that discrimination and the legacy of discrimination, not preference or income, continues to be the leading cause of segregated housing markets in the 1980s.

The legacy of the past, particularly the actions of local governments, is described by Yale Rabin in "The Roots of Segregation in the Eighties: The Role of Local Government Actions." Urban renewal projects, the location of highways, clearance of black neighborhoods, and other actions all had the effect of reinforcing segregated housing patterns. Failure to take any actions on the part of these same governments in the 1970s and 1980s to reverse the patterns that they helped create in earlier years is a major contributor to the continuation of segregated housing today. This theme is further expanded by Goering and Lief in "The Implementation of the Federal Mandate for Fair Housing." The federal government, particularly through FHA and HUD, has a long history of discriminatory housing practices. Even more damaging, the failure to take affirmative acton in the 1970s and 1980s to implement the legislation and the regulations that they are mandated to enforce, helps perpetuate systematized patterns of segregation that are found in the 1980s. More than any other factor, Goering and Lief show that the failure to act on the part of the federal government remains the key and most damaging aspect of continued segregation in this decade.

How segregated are America's cities in the 1980s? John Farley found, in his analysis of the 1980 census, that blacks were only slightly less segregated in 1980 than they were in 1970. These patterns are likely repeated in the six years since the census was taken. Even the modest improvement may be an illusion. The census data show that blacks are much more likely to be found in the suburbs than a decade ago. Furthermore, it appears that some modest integration has occurred in the suburbs. On closer examination, the data reveal that most blacks who live in the suburbs live in predominantly black neighborhoods, or in neighborhoods that changed rapidly from 1970 to 1980. There is no evidence that a significant number of neighborhoods are truly integrated, and not merely showing a racial mix in a transition period from white to black. Thomas Clark has shown that black suburbanization usually repeats patterns of resegregation in lower-density areas. In most cases, however, black enclaves have merely extended beyond their central city confines to include inner-ring suburbs as well.

Woolbright and Hartmann, in their article "The New Segregation: Asians and Hispanics," found most Asian and Hispanic groups highly segregated as well. Some of this segregation can be attributed to the new arrival status of many of these groups. But Puerto Ricans, for example, have been in the continental United States for a number of decades. While some Asians of high-income status can be found in relative deconcentration in the suburbs, most Asians remain highly segregated in inner-city or suburban ghettos. But neither group seems to face the magnitude of both covert and overt barriers that blacks still encounter in metropolitan America.

While the legacy of the past continues to constrain minorities within segregated housing, a number of the authors in this volume demonstrate that continued discrimination plays an important role as well. Orfield and Fossett found systematic patterns of housing discrimination in Chicago, while Hansen and James found the same in small cities in Colorado.

Discrimination in the 1980s is certainly more subtle. But coupled with the severe constraints of the past, and the ways that separate neighborhoods have been institutionalized into the housing market, any continued discrimination has extreme effects. Realtors continue to steer their clients into certain areas, and landlords refuse to rent on the basis of race. Lenders do not make critical loans in areas that are no longer officially redlined, but nevertheless loans are not made in certain areas. The combined effect of these actions is to reinforce both old patterns of housing choice, and perceptions of the inevitability of segregated neighborhoods.

The data in this volume present an irrefutable picture. First, and foremost, segregated housing patterns still persist. They have not much changed in terms of character, and there is little evidence of their gradual disappearance. Enough blacks and other minorities have reached middle-class and upper middle-class status to achieve a modicum of integration over the past two decades. Instead, the patterns of resegregation continue, racial transition being the norm. The economic progress of a significant proportion of the minority populations over the past few decades has not brought with it the commensurate changes in housing patterns that one would have expected.

Decades ago, apologists for a segregated housing system argued that better economic times for minorities would bring, as a natural consequence, racial integration. But the powerful combination of past discriminatory actions, and continued discrimination today, prevents what ought to be a "natural" outcome. Perhaps most importantly, the failure to take affirmative actions in the 1970s and 1980s to dismantle the segregated housing market is the most obvious problem of all. Patterns that have been institutionalized over many decades are not going to disappear on their own. Unless specific actions are taken by governments at all levels to dismantle the existing segregated housing system, there is little hope for change in the near future. The private market in housing has not existed for at least half a century. It is a highly regulated and subsidized industry. Private market forces are merely a reflection of what the government has done in the past, or what it fails to do in the present.

What governments are doing currently tends to be counterproductive. Most subsidized housing programs have been slowly but effectively phased out over the past six years. Those that do remain make no attempt in any meaningful way to integrate existing units. At the same time, the slow but steady processes of inner-city decay, and now inner-suburb decay, continue unabated. These declining and deteriorating bases of black and other minority population continue to destablize housing markets throughout the United States. As has been the case for decades, blacks and other minorities continue to seek housing in new environments, only to find transition and ultimately neighborhood decay at the other end of a series of what seem to be perpetual moves.

The pathology of transition continues to erode, and often to devastate neighborhoods. The subsidies to high-income housing through the income tax structure, and subsidies to outer suburbs' physical infrastructures continue. The federal government has never wavered from its program of high-income and high middle-income household subsidies. With it, of course, they have subsidized the transition process. Unfortunately, as has been shown throughout the literature, what "filters" down is not only housing but entire neighborhood packages as well. Schools, streets, and other urban services decline in quality or perceived quality as the transition process accelerates. As this phenomenon continues to affect inner cities and inner suburbs, the impetus for white flight from particular neighborhoods and white avoidance of those neighborhoods continues as well. These processes, of course, are themselves abetted by the failure of realtors to show whites housing in an integrated neighborhood, and the failure of lenders to act quickly and fully on loan applications in these areas.

At the same time, subsidies to selected gentrifying neighborhoods have accompanied the general decay in other parts of the cities throughout the 1970s and 1980s. But, as has been shown in the literature, gentrification affects a very small proportion of neighborhoods. Even where some modest integration is achieved in redevelopment areas, the consequences often include displaced low-income minority populations again seeking alternate housing in transition areas. It is a very old pattern that has taken new forms in the 1970s and 1980s, but with the same results. The middle class and upper-middle class continued to be subsidized in the housing market, creating dislocation and transition in other neighborhoods. The net result, of course, is continued segregation and resegregation throughout metropolitan areas.

Positive government action has been rare, even when the national will has been turned to civil rights issues. But in the 1980s, the national attention has been concentrated on many issues. War has been declared on organized crime, on pornography, and against drug use. Abortion continues to be a hotly contested issue, as it was during the 1970s. Prayer in public schools has absorbed much public attention. Tax reform has been a key focus in the 1980s, along with a stronger national defense. All of these make up a social and political agenda that rarely, if ever, includes civil rights as a component.

Indeed, concern with civil rights issues has all but vanished from public debate. The current "crisis" agenda has been determined by the social and political, and often the religious right. Affirmative action, in any form, is not only neglected but is looked upon in the current social context with scorn.

The persistence of housing segregation in the 1980s, and the persistent forms of discrimination that are demonstrated in this volume, require affirmative action. But the prospects are quite dim. Housing audits, investigative actions initiated by the federal government, increased efforts to desegregate subsidized housing, increased development funds for integrated neighborhoods, and a variety of other actions are needed to enforce existing legislation and regulations, and at the same time effectively dismantle a well-established system. It is clear that such actions are not going to take place in the current political climate.

It is useless to argue whether or not housing is slightly less segregated in the 1980s than it was two decades ago. Significant progress has not been made. The vast majority of white Americans, the vast majority of black Americans, and other minorities all still live apart from one another. It has been demonstrated over and over again that the social and economic costs, to say nothing of the moral costs, of a segregated society are extraordinarily high. It is clear from the vantage point of the mid-1980s that it is a cost that American society will continue to bear for a long time to come.

1

Choosing Neighbors and Neighborhoods: The Role of Race in Housing Preference

JOE T. DARDEN

IN THE 1980s,America's neighborhoods remain racially segregated. The causes of neighborhood segregation continue to be debated. At least three different theories have been presented to explain neighborhood segregation. First, class theory states that racial groups are distributed unequally by income, education, and occupation. Since neighborhoods are located in different parts of metropolitan areas by quality, cost of housing, and other amenities, racial segregation of neighborhoods occurs (Galster, 1982, p. 40).

Second, discrimination theory states that nonwhite racial groups are denied equal access to housing in white neighborhoods by a combination of such practices as white refusal to sell or rent, racial steering by white real estate brokers and other prohibitive actions by financial institutions, builders, and state, local, and federal governmental agencies (Darden, 1973).

Third, voluntary segregation or preference theory suggests that whites and nonwhites prefer to live in separate neighborhoods with members of their own race. As a result, blacks and whites voluntarily segregate themselves into racially homogeneous neighborhoods (Galster, 1982, p. 40).

The purpose of this chapter is to assess the validity of each of these theories in explaining racially segregated neighborhoods. However, the third theory will be examined more thoroughly since relatively fewer studies have been conducted of neighborhood preference and its influence on segregated neighborhoods.

ASSESSMENT OF THE CLASS THEORY OF NEIGHBORHOOD SEGREGATION

In a controversial book entitled *The Declining Significance of Race,* Wilson speculated that class factors, that is, economic ability, may be more important than race in explaining where people live (Wilson, 1978, p. 141). The class theory, however, has been subjected to several empirical tests and the results have consistently revealed that class is *not* the major reason for black neighborhood segregation. The conclusions of a few of those studies are presented here.

Farley (1977) analyzed 1970 census data for 29 urbanized areas using the index of dissimilarity. He was interested in answering the question, "To what extent are blacks in a given social group residentially segregated from whites in that same social group?" Farley measured the residential segregation between blacks and whites controlling for social class. His findings were that black residential segregation in the United States is not primarily a function of racial differences in socioeconomic status. He concluded that the levels of racial residential segregation did not vary spatially by class.

Massey's (1979) analysis of 29 urbanized areas revealed little or no relationship between socioeconomic variables and black-white residential segregation. The high degree of residential segregation between blacks and whites could not be accounted for by socioeconomic factors alone.

Finally, Farley (1983), using 1980 census data on blacks in metropolitan St. Louis, Missouri, found that the segregated distribution pattern of blacks was not strongly related to the cost of housing. Socioeconomic differentials between blacks and whites can account for less than 15% of the segregation among suburbs and less than 25% of the segregation in the central city of St. Louis. In short, if blacks and whites were residentially distributed only according to their socioeconomic status, with no independent effect of race, St. Louis would be a very integrated area, in sharp contrast to the highly segregated area that exists in reality.

Further empirical evidence showing the invalidity of the class theory is presented in Tables 1.1, 1.2, and 1.3. Although the data are on the Kansas City, Missouri Metropolitan Area, it is typical of several metropolitan areas in the United States. Table 1.1 clearly shows that given the same occupation, most blacks and whites do not live in the same neighborhoods. Occupational status, therefore, does not explain the high level of residential segregation between blacks and whites, and black professionals are only slightly less segregated from white professionals than black laborers are from white laborers.

Similar to black and white workers, most blacks and whites with the same level of education live in separate neighborhoods. Segregation is high regardless of educational level (Table 1.2).

Segregation is also uniformly high between blacks and whites with equal incomes. Blacks and whites earning $30,000 per year are no less segregated from each other than blacks and whites earning $5,000 per year. The level of segregation for both income groups was 82% in 1980 (Table 1.3).

In sum, most empirical studies show that blacks and whites in poverty usually live in separate neighborhoods and so do affluent blacks and whites (Farley, 1977; Simkus, 1978; Straszheim, 1974). If families were distributed over neighborhoods on the basis of class instead of race, most neighborhoods would contain numerous blacks and whites, and residential segregation in cities and their suburbs would be low (Farley & Colasanto, 1980; Langendorf, 1969).

Thus neighborhood segregation between blacks and whites does not occur merely because blacks are poorer, less educated, or in lower-status jobs. The "nature of the beast" is race, not class. Therefore, black socioeconomic mobility does not guarantee freedom of spatial mobility, that is, freedom to move into the neighborhood of choice subject only to ability to pay. Thus for more blacks to have incomes equal to whites would not in and of itself reduce neighborhood segregation by race.

ASSESSMENT OF THE DISCRIMINATION THEORY OF NEIGHBORHOOD SEGREGATION

Although blatant discriminatory practices in housing are not as common today as they were prior to the passage of the Fair Housing Act of 1968 (1976), a great deal of research documents the validity of the discrimination theory as an explanation for neighborhood segregation.

TABLE 1.1
Residential Segregation Between Blacks and Whites with the Same Occupations, Kansas City, MO, SMSA, 1980

Occupations	*Number by Race* White	*Number by Race* Black	*Dissimilarity Index* Black vs. White
Executive, Administrative, and Managerial	25,986	2,908	76.0
Professional Speciality	25,680	3,903	72.0
Technicians and Sales	30,016	3,736	75.1
Administrative Support including Clerical	45,984	10,026	77.1
Service	24,299	12,338	82.3
Precision Production, Craft, and Repair	24,852	3,601	82.1
Operators, Fabricators, and Laborers	29,591	9,780	80.1

SOURCE: Computed by the author from data obtained from U.S. Bureau of the Census, 1983.

In general terms, racial discrimination in housing exists whenever individuals, in this case blacks, are prevented from obtaining the housing they want in the location they prefer for racial reasons. Black individuals are prevented from exercising a purely economic choice through the use of techniques and/or policies designed to avoid selling or renting housing to blacks in a given location. Such discrimination by race hampers the development of integrated neighborhoods, despite improved white attitudes toward blacks and increases in the socio-economic status of blacks (Foley, 1973).

Such discrimination flourishes in the absence of a single real estate market in American cities. All persons cannot make their housing choices entirely according to income and preference. In reality, two housing markets exist, that is, one for blacks and one for whites. It is through the creation and perpetuation of these two separate markets that discrimination in housing and black residential segregation are facilitated. Black residential segregation is the result of a cumulative process involving past and/or present discriminatory practices by the following principal actors: (1) real estate brokers, (2) home builders, (3) financial institutions, (4) the federal government, and (5) state and local governments. Like other urban social problems that have deep roots in history, black residential segregation cannot be totally understood

TABLE 1.2
Residential Segregation Between Blacks and Whites with the Same Level of Education, Kansas City, MO, SMSA, 1980

Educational Level	*Number by Race* *White*	*Black*	*Dissimilarity Index* *Black vs. White*
Elementary			
0-4 years	3,321	2,442	83.5
5-7 years	10,289	4,267	83.9
8 years	20,394	4,242	82.9
Mean			83.4
High School			
1-3 years	45,082	19,121	83.6
4 years	131,590	27,839	79.6
Mean			81.6
College			
1-3 years	60,070	12,3992	74.4
4 years	30,342	2,932	76.4
5 or more years	21,985	2,280	71.5
Mean			74.1

SOURCE: Computed by the author from data obtained from U.S. Bureau of the Census, 1983.

without comprehending the roles each actor has played in causing these patterns.

Real estate brokers have played the role of primary "gate keepers" through whom most information and activity in the housing market must pass. Traditionally, real estate brokers have operated on the assumption that residential segregation is a business necessity and a moral absolute (U.S. Commission on Civil Rights, 1973, p. 3).

The following three discriminatory techniques have been the most commonly used by real estate brokers: (1) overt refusal to show or sell blacks homes in predominantly white areas, (2) the maintenance and control of two or more separate listings, and (3) racial steering.

Prior to the 1968 Fair Housing Act, white real estate brokers simply refused to show or sell blacks homes in predominantly white areas. McEntire (1969, p. 240) cites a San Francisco study in which interviews with representatives of sixty-four real estate firms handling residential properties in most sections of the city led to the conclusion that 80% of the white brokers offered their services to prospective black home

TABLE 1.3
Residential Segregation Between Blacks and Whites with the Same Income, Kansas City, MO, SMSA, 1980

Median Family Income Level	*Number by Race* *White*	*Black*	*Dissimilarity Index Black vs. White*
Less than $2,500	2,121	1,926	86.6
2,500 to 4,999	2,579	2,168	80.8
5,000 to 7,499	4,752	2,457	82.6
7,500 to 9,999	6,167	2,374	85.6
10,000 to 12,499	7,514	2,249	83.6
12,500 to 14,999	7,548	2,399	86.1
15,000 to 17,499	8,270	2,110	85.9
17,500 to 19,999	8,874	1,866	86.9
20,000 to 22,499	9,875	1,705	82.6
22,500 to 24,999	8,521	1,375	86.1
25,000 to 27,499	8,943	1,398	82.4
27,500 to 29,999	6,780	934	82.4
30,000 to 34,999	11,479	1,508	82.4
35,000 to 39,000	7,449	852	84.0
40,000 to 49,999	7,046	1,060	82.1
50,000 to 74,999	4,037	358	88.1
75,000 or more	1,337	109	95.6

SOURCE: Computed by the author from data obtained from U.S. Bureau of the Census, 1983.

buyers either on a restrictive basis or not at all. Helper's (1969, pp. 40-41, 317) survey of ninety brokers in Chicago revealed that 83% of the brokers would never sell to a black in an all-white area. The brokers indicated that they would sell to blacks only in areas where other blacks were already living.

The Civil Rights Act of 1968 and the Supreme Court decision of the same year have made discriminatory behavior by white real estate brokers more difficult. Although black home seekers still occasionally encounter outright refusal of brokers' services, it is no longer common practice for white brokers simply to refuse to sell a particular house to a black home seeker. Discriminatory behavior, however, has not disappeared. The basic means by which real estate brokers can continue to discriminate, despite civil rights legislation and antidiscriminatory court decisions, is through the control of the flow of information about houses that are for sale (Yinger, 1975, p. 30). It is very difficult for law enforcement officials to verify whether blacks and whites are treated equally because of the volume and complexity of this information. Evidence suggests they do not receive equal treatment. Less information is usually available for black customers than for white customers (National Committee Against Discrimination in Housing, 1970, p. 69). A particular system devised by brokers to carry out discrimination within the framework of this information flow involves the maintenance and control of two or more separate listings. The maintenance and control of two or more separate listings became more common as a discriminatory technique after the 1960s.

The most effective, subtle, and widespread discriminatory technique used by white real estate brokers in the 1980s is "racial steering." This is a practice by which a real estate broker directs buyers toward or away from particular houses or neighborhoods according to the buyer's race (Aleinikoff, 1976, p. 809; Openshaw, 1973; Saltman, 1975, pp. 43-45; U.S. Commission on Civil Rights, 1971, pp. 60-61). Black home seekers are steered away from white areas, while whites are directed to them. Conversely, white home seekers are steered away from black areas while blacks are directed to them. Racial steering can be divided into two broad types of conduct: (1) advising customers to purchase homes in particular neighborhoods on the basis of race; and (2) failing to show or inform potential buyers of homes that meet their specifications, on the basis of race.

Real estate brokers have used a variety of reasons to justify racial steering. One reason is based on the longstanding assumption of the housing industry that "majority and minority groups will not willingly share the same residential areas" (McEntire, 1960, p. 2). This assumption results from traditional housing industry values. Until 1968, the Code of Ethics of the real estate industry encouraged racial steering. Thus a white broker may steer white buyers toward white areas in the belief that he or she is satisfying the *preferences* of the home buyer and steer black buyers away from such areas in order to preserve the values of the white areas in which the broker operates (Aleinikoff, 1976, p. 812; Barresi, 1968, p. 60). White brokers argue, therefore, that they are simply fitting into a social structure that itself prefers and maintains racial segregation, and that brokers should not introduce racial integration because it is undesirable and unacceptable to the buyer, seller, and established residents.

Denton's (1970) study of San Francisco clearly showed how minorities were excluded from renting housing outside of established nonwhite neighborhoods. He concluded that the vast majority of apartment owners discriminate, and almost all believe that their white tenants will leave if they rent any of their apartments to minority families.

Widespread discrimination in housing has continued to affect various racial and ethnic groups. However, the most serious and persistent discrimination has been against blacks. It is no accident that blacks are the most residentially segregated and the least suburbanized minority group (Tables 1.4 and 1.5).

Farley, Schuman, Bianchi, Colasanto, and Hatchett (1978) concluded that resistance by whites to housing desegregation has far more explanatory power for continued racial separation of neighborhoods than either the economic status of blacks or the theory that blacks prefer to live in black neighborhoods.

Furthermore, most white Americans are well aware that racial discrimination in housing exists and that some blacks miss out on "good housing" because of discrimination (Pettigrew, 1973, p. 30). Thus considering a black and a white of equal socioeconomic status, the white will probably reside in a better neighborhood with a higher tax structure, better schools, services, and cultural advantages (Villemez, 1980, p. 416). Discrimination results in lower-class, middle-class, and upper-class blacks living in closer proximity on the average than do lower-class whites to the white middle and upper class (Erbe, 1975). Such clustering of blacks regardless of class is often mistakenly attributed (by whites) to black preference for segregated living. Many whites in both the North and the South have a tendency to comfort themselves with the idea that most blacks want to live in black neighborhoods.

ASSESSMENT OF THE PREFERENCE THEORY OF NEIGHBORHOOD SEGREGATION

Despite the overwhelming evidence that the racial discrimination theory is valid, the preference theory has continued to be advanced as the primary reason blacks live in segregated neighborhoods. The assumption of this theory is that blacks do indeed have freedom of spatial mobility (especially after passage of the Fair Housing Act) and those blacks who remain in predominantly black neighborhoods are there by choice (Coleman, 1979, p. 11; Wolf, 1981, pp. 34-39).

It is conceivable that some blacks might prefer to live only with other

TABLE 1.4
Population and Racial Minority Group Segregation in Standard Metropolitan Statistical Areas of Michigan
1980

Part I

SMSA	*Asian Population*	*Native American Population*	*Black Population*	*Hispanic Population*
Ann Arbor	5,631	720	28,323	4,055
Battle Creek	617	626	13,643	2,639
Bay City	289	604	1,023	3,162
Benton Harbor	822	593	24,817	2,088
Detroit	33,257	12,483	890,417	71,589
Flint	2,084	3,013	78,871	8,467
Grand Rapids	3,071	2,479	32,092	13,748
Jackson	522	495	10,840	1,807
Kalamazoo	1,536	1,126	20,887	4,104
Lansing	3,315	2,815	24,882	14,037
Muskegon	428	1,341	19,217	3,965
Saginaw	811	916	35,841	12,353

Part II

Index of Dissimilarity

SMSA	*Asian vs. White*	*Native American vs. White*	*Black vs. White*	*Hispanic vs. White*
Ann Arbor	31.3	24.1	44.5	25.0
Battle Creek	41.5	33.6	61.5	26.0
Bay City	15.2	31.0	52.2	33.4
Benton Harbor	29.8	26.3	82.1	38.5
Detroit	26.1	36.3	85.8	43.6
Flint	25.3	30.0	83.3	33.4
Grand Rapids	27.1	40.8	71.9	48.9
Jackson	26.0	46.1	71.5	39.4
Kalamazoo	22.9	30.7	59.1	29.0
Lansing	33.0	38.6	44.7	37.8
Muskegon	27.2	36.9	63.0	34.2
Saginaw	18.9	44.1	83.0	54.6
Mean	27.0	34.8	66.8	36.9

SOURCE: Computed by the author from data obtained from U.S. Bureau of the Census, 1982.

blacks even if they had total freedom to choose their neighborhood. The explanation for their preference, however, cannot be totally divorced from past and present forces of racism and discrimination (Darden, 1973, p. 64; Goodman & Streitwieser, 1982). Since blacks have never

TABLE 1.5
The Percentage of Each Racial Group's Population
That Is Living in the Suburbs of Michigan's SMSA,
1980

Suburbs Ranked by Suburban Population	*Whites*	*Asians*	*Indians*	*Hispanics*	*Blacks*	*SMSA Population Suburbanized*
Benton Harbor	98.7*	99.1*	93.6*	93.3*	48.9	91.4
Battle Creek	84.3*	80.6	67.9	76.2	40.3	80.9
Muskegon	80.2*	67.8	75.8	69.3	54.5	77.3
Jackson	76.3*	76.1*	57.9	55.3	43.3	73.8
Detroit	87.7*	80.1*	72.6*	59.5	14.8	72.4
Kalamazoo	74.3*	59.4	70.4	63.8	40.5	71.4
Grand Rapids	73.5*	63.2	49.3	58.2	10.9	69.8
Flint	79.3*	69.6*	67.8	53.1	16.2	69.4
Saginaw	75.6*	77.2*	56.4	42.8	23.1	66.0
Bay City	66.1*	76.1*	60.9	38.4	27.2	65.3
Lansing	75.6*	75.8*	50.2	41.3	26.9	63.7
Ann Arbor	59.8*	31.5	62.6*	44.5	64.7*	59.5
Average	77.6	71.4	65.5	58.0	34.3	71.7

SOURCE: Computed by the author from data obtained from U.S. Department of Commerce. Bureau of the Census.
* = higher than the percentage of the total SMSA population that is living in the suburbs.

had the total freedom to live in any neighborhood within cities and suburbs, the influence of personal preference cannot be adequately measured. In order for black housing choices to represent more nearly "real" choice rather than acceptance of a needed commodity, blacks must be free of constraints. Since blacks have never been free of constraints, the influence of black housing preference on segregation remains hypothetical at best.

The validity of this theory rests on three basic premises: (1) Most blacks actually prefer to live in all or predominantly black neighborhoods; (2) Most blacks are actually realizing their preference, that is, most are living in the preferred neighborhood; (3) Since most blacks and whites are living in the neighborhood of their choice, both are equally satisfied with their neighborhood conditions and services. If each premise can be verified, then the preference theory of neighborhood segregation may be valid. If, on the other hand, the three premises cannot be varified, then the preference theory must be seriously questioned.

DO BLACKS PREFER PREDOMINANTLY BLACK NEIGHBORHOODS?

Within the hypothetical context, the black preference theory has been assessed employing surveys of black attitudes toward housing and race. In a 1968 study of the attitudes of blacks in fifteen cities, the black preference for nonsegregated neighborhoods was revealed (Campbell & Schuman, 1968). Blacks were asked, "Would you personally prefer to live in a neighborhood with all Negroes (blacks), mostly Negroes, mostly whites, or a neighborhood that is mixed half and half?" Nearly half (48%) said they preferred a mixed neighborhood; 37% said it makes no difference; only 8% said they preferred an all-Negro neighborhood; and another 5% said they preferred a mostly Negro neighborhood.

One of the most extensive assessments of black preferences was done by Pettigrew (1973). He presented a variety of survey findings concerned with housing and neighborhood choice. He concluded that when blacks have a true option between an all-black neighborhood and a mixed neighborhood, black respondents overwhelmingly favored the latter.

In 1969, 63% of even low-income black Southerners favored desegregated neighborhoods. Those who favored racially mixed areas stated that they did so for positive reasons of racial harmony even more than the obvious advantages of good neighborhoods. Similarly, those who preferred all-black neighborhoods made it clear that their reasons were related to a desire to avoid interracial tension and strife (Pettigrew, 1973, pp. 43-51). In other words, their preference was strongly influenced by the perceived reactions of whites to black residency in the neighborhood. Young respondents in the North were the segment strongest in their desire for all-black neighborhoods—though this still came to only 17% compared to 77% preferring mixed neighborhoods (Pettigrew, 1973, p. 52). Most blacks also did not prefer all-white neighborhoods.

The black preference for mixed neighborhoods has been confirmed by other studies. The 1976 Detroit Area Study investigated residential segregation through a survey of attitudes about race and housing (Bianchi, 1976). Respondents were householders selected in a two-stage, area probability sample of the Detroit metropolitan area. The race of the interviewer was controlled so that black respondents were always interviewed by blacks and white by whites. A total of 734 whites and 400 blacks were interviewed (Farley & Colasanto, 1980, p. 624). The response rate was 78% for whites and 71% for blacks. The survey revealed that 31% of the blacks in the Detroit Standard Metropolitan Statistical Area did not prefer to live in all black neighborhoods. At the other end of the continuum, 62% of the blacks in the metropolitan area did not prefer to move into an all white neighborhood because of perceived white hostility. However, almost all blacks were willing to live in neighborhoods that were racially mixed. The truly preferred neighborhood is one that is 50% white and 50% black (Taylor, 1979, p. 34). Galster (1982) examined two data bases on individual household interviews. The data bases contained information on the occupants, the dwellings, and the associated neighborhoods. The two surveys were conducted in St. Louis in 1967 and in Wooster, Ohio in 1975. Similar statistical estimates were made for both data sets, thus allowing one to compare results from a large city such as St. Louis, which was undergoing rapid expansion of a black ghetto, to those of a small city with few widely diffused blacks. Using multiple-regression analysis, Galster examined the housing price gradient for white and black households across areas of different racial mixtures. He concluded that St. Louis black renters are indifferent to racial composition whereas black owners in general have (1) a predilection for as many white neighborhoods as possible due to the value placed on integrated neighborhoods per se but the perceived isolation from the black cultural milieu associated with extremely white areas reduces this preference somewhat, and (2) blacks have an aversion to areas with increasing percentages of blacks once those neighborhoods were already predominantly black, since such areas represent lower status and certain neighborhood disadvantages. Among those disadvantages are unequal schools, recreational facilities, police protection, garbage collection, health care facilities, and access to mortgage and home improvement loans (Darden, 1985; Schuman & Gruenberg, 1972).

Galster's (1982) findings are consistent with previous studies involving surveys of black preferences. They indicate a black preference for mixed or half-black half-white neighborhoods and the rejection of all-black and all-white ones (Davis & Casetti, 1978; Farley et al., 1978; Schuman & Hatchett, 1974; Taylor, 1981).

Furthermore, compared to other factors such as cost and quality of housing, racial composition of the neighborhood is a relatively unimportant consideration in neighborhood choice (Taylor, 1981, p. 269).

All households (white and nonwhite) look for features that constitute good housing and a good neighborhood with an associated high level of quality services (Foley, 1973, p. 86). The desire for high quality schools has been consistently emphasized (Billingsley, 1968; Foley, 1973; Mack, 1968; Schermer & Leven, 1968; Taylor, 1981; Watts et al., 1964). Other services include police protection, access to recreation, work location, health care, public transportation, and shopping. Both majority and minority households are driven by socioecomonic status and the life cycle. With the exception of racial composition (most whites do not prefer mixed neighborhoods), the preferences of minorities for residential locations (all things being equal) do not differ significantly from those of the white majority (Foley, 1973, p. 90). Blacks with young children, for example, like whites, prefer neighborhoods with the better schools and other amenities of housing in the suburbs. Market surveys conducted among middle-income and moderate-income black households indicate a strong preference for detached houses, individual lots, and other features that are more characteristic of suburbia than the central city (Schermer & Leven, 1968, p. 20). Yet blacks are extremely underrepresented in the suburbs.

In 1980, about three-quarters of the white population and four-fifths of the black population were located in metropolitan areas (i.e., central cities and suburbs). However, almost one-half of the white population lived in neighborhoods in the suburbs compared to less than one-quarter of the black population. On the other hand, more than half of the black population (57.8%) lived in neighborhoods in central cities, compared to exactly one-quarter of the white population (O'Hare, 1982, p. 20). The fact that few blacks live in the suburbs does not necessarily mean that neighborhoods in the suburbs are not preferred. It may mean that ultimate wants are not being satisfied.

Black housing choice, like white housing choice, is influenced by life cycle stages (Mann, 1973; Rossi, 1955). There are at least three stages, namely, the initial or baseline stage, the incremental change stage, and the approximation of the ideal stage (Michelson, 1977, p. 35). The initial baseline stage is the first stage either of the formation of a family or in a family's arrival within a given metropolitan area. Thus the most important neighborhood locational factors in this stage are housing cost, social contacts, and access to the workplace.

During the incremental change stage, the household focuses on the "model home," that is, the ideal place to live. Most studies on housing in North America have indicated that for all racial, ethnic, and social groups, the ideal is the self-contained single-family house, in low-density areas (Hinshaw & Allott, 1972; Lansing, 1966). Yet large segments of the population, especially blacks, do not live in such residences. Apparently there is a separation of the black ideal from the reality. Thus stage 2 can be seen to include a number of strains, stresses, and resolutions from the baseline stage continuing either indefinitely or until stage 3 is reached (Michelson, 1977, p. 35).

Stage 3, the approximation of the ideal, reflects those elements of a neighborhood that people consider most important. Those most likely to be experiencing this stage are households free of constraints. Only when constraints are removed do responses truly reflect housing and neighborhood choice (Michelson, 1977, p. 68). In order for responses to reflect choice, people must be free of constraints. Otherwise, the responses may be biased by current constraints and not indicate what the same people might do on other occasions with no constraints. In the United States, which is a socially stratified society with racist attitudes prevalent, the only households actually free of constraints are affluent white households. Whereas white neighborhood location can be assessed in terms of preference, black neighborhood location must be measured in terms of constraints that have prevented the fulfillment of preferences (Frey, 1984, p. 804). We turn now to the second question, that is, are most blacks actually realizing their preference and living in the *preferred* neighborhood? Before addressing this second question, however, it may be helpful to reiterate the findings of research on the first premise. Most blacks prefer mixed neighborhoods as opposed to all or predominantly black ones. The first premise therefore is simply not true. Racial composition of neighborhood aside, most blacks (like whites) prefer neighborhoods with an abundance of good, single-family, high-quality, detached houses on individual lots with high-quality services and other amenities that are more characteristic of the suburbs than the central city. Are black preferences being realized? We explore now the validity of the second premise.

ARE MOST BLACKS LIVING IN THE NEIGHBORHOOD OF CHOICE?

Given the high level of black residential segregation in 1980 (Table 1.6), it is obvious that most blacks do not live in racially mixed neighborhoods. Within 318 metropolitan areas in the United States in 1980, the majority of blacks (59%) would have to change their place of residence in order to bring about desegregated neighborhoods (Table 1.6). Blacks were more likely *not* to live in mixed neighborhoods in

TABLE 1.6
Mean Indexes of Dissimilarity for SMSAs in Michigan and in National Geographic Regions, 1980

Area	*Mean Index*	*N*
Michigan	66.8	(12)
New England	50.2	(29)
Middle Atlantic	66.6	(34)
East North Central	67.0	(58)
West North Central	55.3	(26)
Southern Atlantic	59.6	(58)
East South Central	61.6	(20)
West South Central	59.7	(41)
Mountain	46.9	(17)
Pacific	50.9	(35)
Total Regions	59.0	(318)

SOURCE: For the mean index on Michigan, see Darden, 1985, chap. 3; for regional indexes, see Jakubs, 1984.

metropolitan areas in Michigan, other East North Central states and in the Middle Atlantic region. Instead, most blacks within metropolitan areas are restricted to predominantly black neighborhoods within central cities. One should not assume, however, that the overwhelming concentration of blacks in central cities is the preferred choice. Were it not for racial constraints, the black pattern of neighborhood selection would follow very closely the life cycle pattern found among whites.

During the 1950s and 1960s, the white population followed a fairly well-established life course pattern of city to suburb destination selection. The black population, on the other hand, displayed extremely low levels of suburban selection at all ages (Frey, 1984, p. 805). Available evidence suggests that the age-related destination selection patterns of whites during the 1950-1970 period were tied to city to suburb location preferences associated with traditional family life cycle stages (Frey, 1978, 1984; Frey & Kobrin, 1982; Glick & Long, 1976; Johnston, 1971). While the black population also participated in the postwar baby boom and probably would have preferred the suburbs during the child rearing portions of the life course, its low level of suburban selection was due to racial constraints rather than preference (Frey, 1984, p. 805). Evidence shows that the extremely low levels of suburban selection that characterized the black population, in the aggregate, did not vary appreciably across age categories (Frey, 1983) or across any other socioeconomic measures that were related to white suburban selection during the postwar years (Farley, 1976; Frey, 1978, 1984; Nelson, 1980).

As a consequence of constraints caused by discrimination in housing, several cohorts of black movers were directed away from the suburbs at all stages of the life course (Clark, 1979; Foley, 1973; Frey, 1984; Taeuber, 1975). This ensured that the central cities and not the suburbs would become the predominant place of residence for these black cohort members during the older, relatively stable (nonmobile) years of their life (Frey, 1984). The factor was constraint, not preference. The second premise is not valid since most blacks are not living in the neighborhood of choice due to the constraint of discrimination in housing. Because the neighborhood preferences of most blacks are not realized due to racial constraints it is expected that a higher percentage of blacks compared to whites will be unsatisfied with the conditions and services in their neighborhood. We look now at the final premise. Are blacks as satisfied as whites with neighborhood conditions and services?

IS THERE A RACIAL DIFFERENCE IN NEIGHBORHOOD DISSATISFACTION?

Past research suggests that dissatisfaction with the neighborhood is particularly widespread among blacks and other minorities (Hinshaw & Allott, 1972; Kase & Harburg, 1972). Nathanson's (1974) study of a predominantly black neighborhood in Baltimore also suggests that a substantial level of dissatisfaction with local conditions exists. Of the respondents, 67% were unfavorable toward their neighborhood. Schuman and Gruenberg (1972) also found greater black dissatisfaction and attributed it to undesirable objective characteristics of predominantly black neighborhoods such as poorer services. Newman and Duncan (1979, p. 163) found that blacks suffer more than any other demographic group from neighborhood problems of congestion and crime. People of all types see predominantly black areas (when combined with low income) as less desirable places to live. The characteristics of the conditions and poorer services make people, both black and white, low income and high income, adverse to living there (Stipak & Hensler, 1983, p. 319). Where black and white respondents' attitudes have been compared, blacks have consistently been found to be more dissatisfied with neighborhood conditions (Levine, 1972; U.S. Department of Housing and Urban Development, 1978).

The Gallup Poll published in 1982 found that 72% of whites were satisfied with their present housing compared to only 49% of blacks. In total, 68% of whites said they were satisfied with their community as a place to live compared to 57% of blacks (Gallup, 1982, pp. 17-18).

Residence in racially mixed areas in the suburbs, however, seems to nullify dissatisfaction differences. Schuman and Gruenberg (1972) found that dissatisfaction levels virtually diminished between blacks and whites living in racially mixed neighborhoods, with a suburban location having an overall positive effect.

Cutter (1982) examined only suburban black and white homeowners. She concluded that the race of the respondent had no effect in differentiating the satisfaction levels. In fact, black homeowners were slightly more positive in their assessments of neighborhood services and problems than were white homeowners. Cutter's (1982) findings resubstantiated the conclusions of Schuman and Gruenberg (1972). Both studies conclude that it is the racial character of the neighborhood, and not the race of the respondent per se, that influences satisfaction levels. It is clear that the issue of racial differences in neighborhood satisfaction needs further study.

DATA AND METHOD

Data has been provided by the U.S. Bureau of the Census in which racial differences in neighborhood satisfaction may be assessed. Statistics on indicators of housing and neighborhood quality from the 1981 Annual Housing Survey for the United States form the basis of the data that follow (U.S. Department of Commerce, 1983). The data are based on a sample of 60,421 housing units located throughout the United States. Satisfaction differences were assessed separately for blacks and whites in (1) metropolitan areas, (2) central cities, and (3) suburbs.

In order to compare black and white satisfaction levels, the raw data on the characteristics for owner- and renter-occupied housing units with black householders and Spanish-origin households were subtracted from the characteristics for total owner- and renter-occupied housing units. The nonblack, non-Spanish-origin category that resulted may, for all practical purposes, be considered white non-Spanish-origin homeowners and renters. The percentage of black homeowners experiencing unsatisfactory neighborhood conditions and unsatisfactory services compared to all-black homeowners was computed. The same computations were done for nonblack, non-Spanish-origin households. Simple ratios were then computed between the percentage black householders and non-black, non-Spanish-origin householders experiencing unsatisfactory neighborhood conditions and unsatisfactory services. The results are presented in Tables 1.7, 1.8, and 1.9. A ratio of 1.0 means that blacks and whites have equal levels of dissatisfaction.

RESULTS

As Tables 1.7, 1.8, and 1.9 suggest, a higher percentage of black homeowners compared to whites were experiencing 1 or more unsatisfactory neighborhood conditions in metropolitan areas and central cities. Whether the condition was odors, abandoned buildings, crime, or bad streets, a higher percentage of black compared to white homeowners reported that the condition was unsatisfactory. Black homeowners were most dissatisfied with street noise and crime, as were white homeowners. The greatest racial difference in dissatisfaction was with abandoned buildings in the neighborhood. Black homeowners were 4.3 times more likely (than white homeowners) to report dissatisfaction with abandoned buildings in the neighborhood in SMSAs and 3.5 times in central cities.

A higher percentage of renters compared to homeowners reported experiencing 1 or more unsatisfactory neighborhood conditions. Renters were more dissatisfied than homeowners, whether black or white. Like homeowners, the highest percentage of renters reported dissatisfaction with street noise and crime. Indeed a higher percentage of white than black renters were unsatisfied with street noise in the neighborhood. Black renters were 3.5 times more likely (than white renters) to report dissatisfaction with abandoned buildings in the neighborhood in SMSAs and 3.2 times in central cities.

Tables 1.7 and 1.8 indicate the percentage of blacks and whites experiencing unsatisfactory neighborhood services. Whether the service was police protection, recreation or health care facilities, public transportation, access to shopping facilities, or the quality of elementary schools, a higher percentage of black compared to white homeowners reported dissatisfaction with the service. Black and white homeowners were most dissatisfied with recreation facilities. The greatest racial difference was with a lack of access to shopping facilities.

Like homeowners, black and white renters were most dissatisfied with recreation facilities. Also consistent with homeowners, the greatest

TABLE 1.7
Percentage of Blacks and Non-Hispanic Whites Experiencing Unsatisfactory Neighborhood Conditions and Services in U.S. SMSAs, 1981

Neighborhood Conditions	*Black Homeowners*	*Non-Hispanic White Homeowners*	*Ratio*	*Black Renters*	*Non-Hispanic White Renters*	*Ratio*
Percentage Unsatisfied with 1 or more Conditions	68.6	58.3	1.2	77.7	71.0	1.1
Percentage Unsatisfied with:						
Odors, Smoke, Gas	7.9	7.5	1.0	9.9	8.2	1.2
Trash, Litter, Garbage	24.7	12.6	1.9	31.6	16.2	1.9
Abandoned Buildings	18.7	4.3	4.3	25.9	7.3	3.5
Crime	31.2	23.8	1.3	36.8	28.9	1.3
Street Noise	34.8	31.4	1.1	41.2	43.6	0.9
Bad Street	19.0	14.9	1.2	19.5	13.8	1.4
Commercial/Residential Mix	10.8	15.6	0.7	34.6	35.6	0.9
Unsatisfactory Services						
Percentage Unsatisfied with 1 or more Services	40.0	27.7	1.4	43.2	27.0	1.6
Percentage Unsatisfied with:						
Police	14.9	8.6	1.7	18.4	9.3	1.9
Recreation	29.6	18.9	1.5	30.6	18.6	1.6
Health Care	13.4	8.9	1.5	12.2	7.1	1.7
Public Transportation	9.1	6.6	1.3	8.8	6.6	1.3
Shopping	16.2	8.7	1.8	14.1	6.9	2.0
Elementary Schools	6.7	5.2	1.3	5.7	4.6	1.2

SOURCE: Computed by the author from data obtained from U.S. Department of Commerce. Bureau of the Census. Indicators of Housing and Neighborhood Quality by Financial Characteristics for the United States and Regions: 1981: *Annual Housing Survey* 1981, part B. (Washington, D.C.: Government Printing Office, 1983).
NOTE: A ratio of 1.0 = equality.

TABLE 1.8
The Percentage of Blacks and Non-Hispanic Whites Experiencing Unsatisfactory Neighborhood Conditions and Services in U.S. Central Cities, 1981

Neighborhood Condition	*Black Homeowners*	*Non-Hispanic White Homeowners*	*Ratio*	*Black Renters*	*Non-Hispanic White Renters*	*Ratio*
Percentage Unsatisfied with 1 or more Conditions	71.0	63.7	1.1	79.1	75.7	1.1
Percentage Unsatisfied with:						
Odors, Smoke, Gas	8.5	8.3	1.0	10.4	9.0	1.2
Trash, Litter, Garbage	25.6	16.0	1.6	34.3	19.2	1.8
Abandoned Buildings	19.8	5.6	3.5	28.7	9.1	3.2
Crime	35.4	31.9	1.1	39.7	36.3	1.1
Street Noise	35.7	35.7	1.0	43.2	48.0	0.9
Bad Streets	19.1	14.4	1.3	20.2	13.5	1.5
Commercial/Residential Mix	23.2	20.4	1.1	36.4	41.8	0.9
Unsatisfactory Services						
Percentage Unsatisfied with 1 or more Services	40.1	25.6	1.6	45.5	29.0	1.4
Percentage Unsatisfied with:						
Police	16.4	9.9	1.6	20.8	11.5	1.8
Recreation	28.4	17.0	1.7	32.1	20.3	1.6
Health Care	13.5	5.5	2.5	12.3	5.5	2.2
Public Transportation	9.6	7.3	1.3	9.0	7.4	1.2
Shopping	17.1	6.5	2.6	14.7	6.3	2.3
Elementary Schools	8.0	7.2	1.1	6.5	6.1	1.1

SOURCE: Computed by the author from data obtained from U.S. Department of Commerce. Bureau of the Census, 1983.
NOTE: A Ratio of 1.0 = equality.

racial difference was in a lack of access to shopping facilities. Twice as many black compared to white renters reported dissatisfaction with this service.

Table 1.9 shows that suburban black and white homeowners and renters have a lower level of dissatisfaction than homeowners and renters in central cities and SMSAs. Owner occupancy in the suburbs clearly reduces the level of dissatisfaction with neighborhood conditions and services. However, a racial difference gap in dissatisfaction remains. Similar to black homeowners and renters in central cities and SMSAs, black homeowners and renters in the suburbs were most dissatisfied with street noise, as were white homeowners and renters. Crime, trash, litter, and garbage were the other conditions receiving a high percentage of responses among black homeowners. These conditions, combined with commercial/residential mix, were the primary concerns of black renters. Finally, the greatest racial difference in dissatisfaction was with abandoned buildings in the neighborhood. Black suburban homeowners were 4.4 times more likely and black renters were 2.9 times more likely than white homeowners and renters to be dissatisfied with abandoned buildings in the neighborhood (Table 1.9).

Black homeowners and renters were most dissatisfied with recreation facilities, as were white homeowners and renters. Dissatisfaction with recreation and shopping facilities was also where the greatest racial difference occurred.

In sum, the annual housing survey data suggest that homeownership in the suburbs does reduce black and white levels of dissatisfaction. However, some racial difference in dissatisfaction remains with neighborhood conditions and neighborhood services regardless of location. The greatest racial difference in dissatisfaction is with abandoned buildings and trash, litter, and garbage in the neighborhood combined with dissatisfaction with shopping and recreation facilities. It should be reiterated, however, that regardless of residential location, that is, whether central city, suburb, or SMSA, black homeowners and black renters are generally more dissatisfied than white homeowners and renters with neighborhood conditions and neighborhood services.

This difference does not necessarily occur because blacks have more rigid criteria than whites for evaluating neighborhoods (Stipak & Hansler, 1983). Rather, it occurs for two reasons. First, blacks are segregated from whites by neighborhood, whether in central cities or suburbs. Second, blacks living in segregated neighborhoods are less able to convert higher social and economic status characteristics into

TABLE 1.9
The Percentage of Blacks and Non-Hispanic Whites Experiencing Unsatisfactory Neighborhood Conditions and Services in U.S. Suburbs, 1981

Neighborhood Conditions	*Black Homeowners*	*Non-Hispanic White Homeowners*	*Ratio*	*Black Renters*	*Non-Hispanic White Renters*	*Ratio*
Percentage Unsatisfied with 1 or more Conditions	63.3	55.6	1.1	70.4	66.3	1.1
Percentage Unsatisfied with:						
Odors, Smoke, Gas	6.7	7.2	0.9	8.4	7.5	1.1
Trash, Litter, Garbage	22.6	11.0	2.1	21.6	13.2	1.6
Abandoned Buildings	16.3	3.7	4.4	15.7	5.4	2.9
Crime	21.8	20.1	1.1	26.1	21.4	1.2
Street Noise	32.9	29.4	1.1	33.6	39.2	0.8
Bad Streets	15.2	18.8	0.8	16.8	14.1	1.2
Commercial/Residential Mix	15.4	13.4	1.2	28.0	29.4	0.9
Unsatisfactory Services						
Percentage Unsatisfied with 1 or more Services	39.7	28.6	1.4	34.9	25.1	1.4
Percentage Unsatisfied with:						
Police	11.6	8.0	1.5	9.9	7.1	1.4
Recreation	32.3	19.7	1.6	25.2	16.9	1.5
Health Care	13.3	10.4	1.3	11.9	8.6	1.4
Public Transportation	8.0	6.3	1.3	8.0	5.8	1.4
Shopping	14.5	9.7	1.5	12.1	7.5	1.6
Elementary Schools	4.1	4.3	0.9	2.8	3.1	0.9

SOURCE: Computed by the author from data obtained from U.S. Department of Commerce. Bureau of the Census, 1983.
NOTE: A ratio of 1.0 = equality.

residential and neighborhood quality because of discrimination in housing (Villemez, 1980). Simply put, racial constraints reduce the neighborhood options available to blacks even when social and economic constraints have been overcome.

CONCLUSION

The evidence presented in this chapter shows that black residential segregation is best explained by exclusion and discrimination motivated by racial prejudice. Economic factors are of minor importance. Since blacks are not an ethnic group in the way in which foreign born families once were, voluntary segregation is unlikely except as a response to intimidation. Thus racial concentration is largely compelled (Wolf, 1981, p. 26). As a result, it remains severe, widespread, unresponsive to economic improvement, and impervious to the assimilative processes that dispersed ethnic groups (Wolf, 1981, p. 26).

Why then does the theory that "blacks prefer to live among their own kind" continue to be advanced? This argument allows some groups to argue for delaying or preventing efforts toward decreasing black residential segregation (Darden, 1973, p. 64). Such a rationale allows one to support a community's efforts to "maintain the ethnic purity of its neighborhood" without racist guilt (see Citizens Commission on Civil Rights, 1983, p. 49; *New York Times*, 1976, April 7, 9).

IMPLICATIONS FOR FUTURE BLACK/WHITE NEIGHBORHOOD PATTERNS

Given the subtle but persistent discrimination in housing, the lack of an effective enforcement mechanism under the Federal Fair Housing Act of 1968, and the current conservative climate in the Congress and the White House, the future of neighborhood racial integration looks bleak.

In most states, individuals who have experienced discrimination have two options: (1) to request that the United States Department of Housing and Urban Development (HUD) attempt conciliation between themselves and the alleged discriminator, and (2) to file a private suit in Federal District Court. The first option has been proven ineffective due to HUD's lack of authority. Under the 1968 Fair Housing Law, HUD can only conciliate complaints of discrimination, and conciliation has proved to be an ineffective remedy.

If reduction in the present high level of residential segregation is to occur, stronger Federal Fair Housing policies must be introduced. Because of the present practice of racial steering by real estate brokers, major policy changes must be implemented for black residential segregation to be substantially reduced in the near future.

A bill entitled the "Fair Housing Amendments Act of 1983," introduced in the Senate in May 1983, is designed to strengthen the 1968 Federal Housing Law (U.S. Senate, 1983). The principal features of the new bill are designed to provide a speedy, fair, and less costly adjudication process. Specifically, the bill provides for an administrative hearing process, using independent administrative law judges appointed by a three-member Presidential Fair Housing Review Commission. If discrimination is proved, equitable relief (including cease and desist orders) as well as compensatory damages may be awarded. Civil penalties of up to $10,000 may be assessed, and reasonable compensation for attorney's fees may be granted to the prevailing party. This authority to grant relief provides the real enforcement power needed for the Federal Fair Housing Act.

William Bradford Reynolds, Assistant Attorney General for Civil Rights, opposes the bill. The bill has been deadlocked since 1983. At the same time, the Reagan administration has been strongly criticized by civil rights groups for not enforcing the current fair housing law. The Reagan administration, in response to charges of lax enforcement, has proposed legislation to allow the government, like lawyers in private suits, to secure monetary damages for victims of discrimination. That proposal has also been deadlocked for years (Herbers, 1986, p. 22).

Fair housing enforcement is at a critical point. Victims can no longer rely on the federal government to enforce the Fair Housing Law. Such efforts are a part of the past. In the future, we can only hope for more and more victims finding people of goodwill to help overcome the housing barrier. The use of the private suit in federal court will be the primary, and perhaps the only, tool.

Such suits are increasingly being filed, resulting in more court decisions in favor of the minority plaintiffs. It is likely that judgments for plaintiffs will continue to increase in the future. Damage awards may increase as well.

Emphasis on the private suit may result in an increase in private fair housing centers. Such centers will become the primary agencies for fair housing enforcement in the future. Established centers, in metropolitan areas as diverse as Los Angeles and York, Pennsylvania, receive and investigate complaints. Center officials send black and white testers or volunteers, who apply for the sale and rental of housing with the sole intention of determining whether there is discrimination on the basis of race (Herbers, 1986, p. 2; Reid, 1984).

The courts have ruled that in cases involving housing discrimination, unlike some other forms of discrimination, plaintiffs do not have to prove intent to discriminate to win a judgment. Once the plaintiffs establish that there was an act of discrimination, the burden shifts to the defendant to prove there was no illegal exclusion (Herbers, 1986, p. 22).

In sum, with no changes in current policy, the future battle over neighborhood racial integration will continue to shift from the federal government to the private sector. The federal courts will be the arena for settlement. The impact of this development on the overall levels of neighborhood integration must await future study. But without a more active role for the federal and state governments, the future looks bleak.

REFERENCES

Aleinikoff, A. (1976). Racial steering: The real estate broker and title VIII. *The Yale Law Journal, 6*, 808-825.

Barresi, C. M. (1968). The role of the real estate agent in residential location. *Sociological Focus, 1*, 58-71.

Bianchi, S. M. (1976). *Sampling report for the 1976 Detroit area study.* Ann Arbor: University of Michigan.

Billingsley, A. (1968). *Black families in white America.* Englewood Cliffs, NJ: Prentice-Hall.

Campbell, A., & Schuman, H. (1968). *Racial attitudes in fifteen cities.* Ann Arbor: Institute for Social Research.

Citizens Commission on Civil Rights. (1983). *A decent home: Report on the continuing failure of the federal government to provide equal housing opportunity.* Washington, DC: Catholic University Law School, Center for National Policy Review.

Clark, T. A. (1979). *Blacks in suburbs: A national perspective.* New Brunswick, NJ: Rutgers University Center for Urban Policy Research.

Coleman, J. (1979). Destructive beliefs and potential policies in school desegregation. In J. W. Smith (Ed.), *Detroit metropolitan city-suburban relations* (pp. 5-12). Dearborn: Henry Ford Community College.

Cutter, S. (1982). Residential satisfaction and the suburban homeowner. *Urban Geography, 3, 4*, 315-327.

Darden, J. T. (1973). *Afro-Americans in Pittsburgh: The residential segregation of a people.* Lexington: D.C. Heath.

Darden, J. T. (1984). Demographic changes 1970-1980: Implications for federal fair housing policy. In *A sheltered crisis: The state of fair housing in the eighties* (pp. 7-30). Washington, DC: U.S. Commission on Civil Rights.

Darden, J. T. (1985). The housing situation of blacks in metropolitan areas of Michigan. In *The state of black Michigan: 1985* (pp. 11-21). East Lansing: Michigan State University.

Davis, D., & Casetti, E. (1978, July). Do black students want to live in integrated socially homogeneous neighborhoods? *Economic Geography, 54*, pp. 197-209.

Erbe, B. M. (1975). Race and socioeconomic segregation. *American Sociological Review, 40*, 801-812.

Farley, J. (1983). *Segregated city, segregated suburbs: Are they products of black-white socioeconomic differentials?* Edwardsville: Southern Illinois University.

Farley, R. (1976). Components of suburban population growth. In B. Schwartz (Ed.), *The changing face of the suburbs* (pp. 3-38). Chicago: University of Chicago Press.

Farley, R. (1977). Residential segregation in urbanized areas of the United States in 1970: An analysis of social class and racial differences. *Demography, 14*, 497-518.

Farley, R., & Colasanto, D. (1980). Racial residential segregation: Is it caused by misinformation about housing costs? *Social Science Quarterly, 61*, 623-637.

Farley, R., Schuman, H., Bianchi, S., Colasanto, D., & Hatchett, S. (1978). Chocolate city, vanilla suburbs: Will the trend toward racially separate communities continue? *Social Science Research, 7*, 319-344.

Foley, D. L. (1973). Institutional and contextual factors affecting the housing choices of minority residents. In A. H. Hawley & V. R. Rock (Eds.), *Segregation in residential areas* (pp. 85-147). Washington, DC: National Academy of Sciences.

Frey, W. H. (1978). Black movement to the suburbs: Potentials and prospects for metropolitan-wide integration. In F. Bean & W. F. Friskie (Eds.), *The demography of racial and ethnic groups* (pp. 79-117). New York: Academic Press.

Frey, W. H. (1983). *Mover destination selectivity and the Chicago suburbanization of metropolitan whites and blacks* (Research Report No. 83-49). Ann Arbor: University of Michigan, Population Studies Center.

Frey, W. H. (1984). Lifecourse migration of metropolitan whites and blacks and the structure of demographic change in large central cities. *American Sociological Review, 49*, 803-827.

Frey, W. H., & Kobrin, F. E. (1982). Changing families and changing mobility: Their impact on the central city. *Demography, 19*, 261-277.

Gallup, G. H. (1982). *The gallup poll: Public opinion 1982.* Wilmington: Scholarly Resources.

Galster, G. C. (1982). Black and white preferences for neighborhood racial composition. *American Real Estate and Urban Economics Association Journal, 10*, 39-67.

Glick, P. C., & Long, L. H. (1976). Family patterns in suburban areas: Recent trends. In B. Schwartz (Ed.), *The changing face of the suburbs* (pp. 39-67). Chicago: University of Chicago Press.

Goodman, J. L., & Streitwieser, M. (1982). *Explaining racial differences in city-to-suburb residential mobility* (Working Paper No. 1384-09). Washington, DC: Urban Institute.

Helper, R. (1969). *Racial policies and practices of real estate brokers.* Minneapolis: University of Minnesota Press.

Herbers, J. (1986, February 16). Use of private suits in housing bias cases in federal courts is increasing. *New York Times*, p. 22.

Hinshaw, M., & Allott, K. (1972, March). Environmental preferences of future housing consumers. *Journal of the American Institute of Planners, 38*, 102-107.

Jakubs, J. (1984, November). Recent racial segregation in U.S. *SMSAs*. Paper presented at the 39th Annual Meeting of the Southeastern Division, Association of American Geographers, Birmingham.

Johnston, R. J. (1971). *Urban residential patterns*. New York: Praeger.

Kase, S., & Harburg, E. (1972). Perceptions of the neighborhood and the desire to move out. *Journal of American Institute of Planners, 38*, 318-324.

Langendorf, R. (1969). Residential desegregation potential. *Journal of the American Institute of Planners, 35*, 90-95.

Levine, D. U. et al. (1972). Are the black poor satisfied with conditions in their neighborhood? *Journal of the American Institute of Planners, 138*, 168-171.

Mack, R. W. (1973). *Our children's burden: Studies of desegregation in nine American communities*. New York: Random House.

Mann, M. (1973). *Workers on the move: The sociology of relocation*. Cambridge: Cambridge University Press.

Massey, D. S. (1979). Effects of socioeconomic factors on the residential segregation of blacks and Spanish Americans in U.S. urbanized areas. *American Sociological Review, 44*, 1015-1022.

McEntire, D. (1969). *Residence and race*. Berkeley: University of California Press.

Michelson, W. (1977). *Environmental choice, human behavior, and residential satisfaction*. New York: Oxford University Press.

National Committee Against Discrimination in Housing. (1970). *Jobs and housing: A study of employment and housing opportunities for racial minorities in suburban areas of the New York metropolitan area* (Interim Report). New York.

Nathanson, C. A. (1974, September). Moving preferences and plans among urban black families. *Journal of The American Institute of Planners*, pp. 353-359.

Nelson, K. P. (1980). Recent suburbanization of blacks: How much, who and where? *Journal of the American Planning Association, 46*, 287-300.

Newman, S. J., & Duncan, G. J. (1979). Residential problems, dissatisfaction, and mobility. *Journal of The American Planning Association, 45*, 154-166.

O'Hare, W. P. (1982). *Blacks on the move: A decade of demographic change*. Washington, DC: Joint Center for Political Studies.

Openshaw, H. (1973). *Race and residence: An analysis of property values in transitional areas, Atlanta, Georgia 1960-1971*. Atlanta: Georgia State University School of Business Administration.

Pettigrew, T. F. (1973). Attitudes on race and housing: A social-psychological view. In A. H. Hawley & V. P. Rock (Eds.), *Segregation in residential areas* (pp. 21-84). Washington, DC: National Academy of Sciences.

Reid, C. E. (1984). The reliability of fair housing audits to detect racial discrimination in rental housing markets. *AREUEA Journal, 12*(1), 86-96.

Rossi, P. (1955). *Why families move*. Glencoe: Free Press.

Saltman, J. (1975). Implementing open housing laws through social action. *Journal of Applied Behavioral Science, 11*, 39-61.

Scherma, G., & Leven, A. J. (1968). *Housing guide to equal opportunity: Affirmative practices for integrated housing*. Washington, DC: Potomac Institute.

Schuman, H., & Gruenberg, B. (1972). Dissatisfaction with city services: Is race an important factor? In H. Hahn (Ed.), *People and politics in urban society* (pp. 369-392). Newbury Park, CA: Sage.

Schuman, H., & Hatchett, S. (1974). *Black racial attitudes.* Ann Arbor: University of Michigan Press, Center for Survey Research.

Simkus, A. (1978). Residential segregation by occupation and race in ten urbanized areas, 1950-1970. *American Sociological Review, 43,* 81-93.

Spear, A. H. (1967). *Black Chicago: The making of a negro ghetto, 1880-1920.* Chicago: University of Chicago Press.

Stipak, B., & Hensler, C. (1983). Effect of neighborhood racial and socioeconomic composition on urban residents' evaluations of their neighborhoods. *Social Indicators Research, 12,* 311-320.

Straszheim, M. R. (1974). Racial discrimination in the urban housing market and its effects on black housing consumption. In G. M. von Furstenberg, B. Harrison, & A. Horowitz (Eds.), *Patterns of racial discrimination* (Vol. 1). Lexington: D. C. Heath.

Taeuber, K. E. (1975). Racial segregation: The persisting dilemma. *Annals of the American Academy of Political and Social Science, 442,* 87-96.

Taylor, G. (1979). Housing, neighborhoods and race relations: Recent survey evidence. *Annals of the American Academy of Political and Social Science, 441,* 26-40.

Taylor, G. (1981). Racial preferences, housing segregation, and the causes of school segregation: Recent evidence from a social survey used in civil litigation. *Review of Public Data Use, 9,* 267-282.

U.S. Bureau of the Census. (1983). *Population and housing summary tape file 4.* Washington, DC: Data User Services.

U.S. Commission on Civil Rights. (1971). *The federal civil rights enforcement effort: One year later.* Washington, DC: Government Printing Office.

U.S. Commission on Civil Rights. (1973). *Understanding fair housing.* Washington, DC: Government Printing Office.

U.S. Department of Commerce, Bureau of the Census. (1983). Indicators of housing neighborhood quality by financial characteristics of the United States and regions, 1981. *Annual Housing Survey,* Part B. Washington, DC: Government Printing Office.

U.S. Department of Housing and Urban Development. (1978). *The HUD survey on the quality of community life: A data book.* Washington, DC: Government Printing Office.

U.S. Senate. (1983). *Senate bill 1220,* 98th Congress, 1st Session.

Villemez, W. J. (1980). Race, class, and neighborhood: Differences in the residential return on individual resources. *Social Forces, 59,* 413-430.

Watts, L. C. et al. (1964). *The middle income negro family faces urban renewal.* Boston: Brandeis University, Research Center of the Florence Heller Graduate School for Advanced Studies in Social Welfare.

Wilson, J. (1978). *The declining significance of race.* Chicago: University of Chicago Press.

Wolfe, E. P. (1981). *Trial and error: The Detroit school segregation case.* Detroit: Wayne State University Press.

Yinger, J. (1975). *An analysis of discrimination by real estate brokers* (Institute for Research on Poverty Discussion Papers). Madison: University of Wisconsin.

2

The Racial Dimension of Urban Housing Markets in the 1980s

JOHN YINGER

TITLE VIII OF THE Civil Rights Act of 1968 explicitly prohibits racial and ethnic discrimination in the rental or sale of housing. This legislation has been in place for over fifteen years. Has it been effective in eliminating discrimination? Extensive, recent evidence provides an unambiguous answer to this question: Racial and ethnic discrimination in housing is still powerful and widespread.

This chapter reviews this evidence and examines the alternatives for public policy. The second section serves as a conceptual overview and introduction to the issues; the third section reviews the recent evidence on the extent of racial prejudice and racial discrimination in urban housing markets; the fourth section reviews the recent evidence on the causes and consequences of this discrimination; and the fifth section discusses the implications of this evidence for government fair housing policy.

CONCEPTUAL BACKGROUND: THE CAUSES AND CONSEQUENCES OF DISCRIMINATION IN HOUSING

DEFINITIONS

Racial attitudes and racial behavior are of course intertwined in the structure of American society. Nevertheless, clear thinking about the

AUTHOR'S NOTE: *I am grateful to George Galster for extensive helpful comments.*

racial dimension of housing markets requires distinctions among several related phenomena, including prejudice and discrimination. Racial *prejudice* is an attitude. In the housing market, it appears as an aversion toward living with or near members of some racial group. Racial *discrimination* in housing is any type of behavior by people or their agents who sell or rent housing that denies one racial group the same access to housing given to other groups. As documented in the third section, one common form of discrimination is withholding information about available apartments or houses from black home seekers. And as we will see, prejudice is a key cause of discrimination.

Another key phenomenon is racial residential *segregation,* which is the physical separation of the residential locations of different racial groups. This segregation is a matter of degree; that is, the separation can be extensive or modest. Moreover, racial segregation can exhibit different patterns. In American cities, racial prejudice and discrimination contribute to a high degree of racial segregation and to a pattern of segregation with blacks concentrated in a central city and whites concentrated in the suburbs. Residential *integration* by race is the converse of segregation, that is, the extent to which racial groups live together.

One final word on definitions: Minority groups in American society are identified by both race and ethnicity. *Race* refers to certain physiological characteristics of people; *ethnicity* refers to cultural characteristics. This chapter focuses on racial discrimination against black Americans. However, much of the analysis and some of the evidence applies to discrimination against other racial or ethnic groups as well. For example, a few studies examine discrimination against Hispanic Americans, who have a common language and religion, but who do not all have the same racial background. Thus the designation "Hispanic" is largely cultural, and Hispanic Americans may have Caucasian, black, American Indian, or mixed racial ancestry. For more on these distinctions, see Simpson and Yinger (1985).

THE CAUSES OF RACIAL DISCRIMINATION IN HOUSING

Two key hypotheses about the causes of discrimination in housing have appeared in the literature.[1] The first hypothesis, called the customer-prejudice hypothesis, states that landlords and real estate brokers who depend on the business of prejudiced whites discriminate against blacks or other minorities in order to avoid alienating their white customers. In other words, the prejudice of white customers gives housing agents an economic incentive to discriminate.

This hypothesis takes different forms in the rental and sales markets. A landlord with many white tenants may refuse to rent to blacks in order to keep his or her white tenants from moving away, which would force the landlord to incur large turnover costs. In the sales market, a real estate broker cultivates contacts in her or his community so that people will come to a real estate broker when they want to buy or sell a house. Brokers in a prejudiced white community may protect their reputation, and hence their ability to attract white customers, by refusing to sell to blacks in that community.

This economic incentive to discriminate disappears once a neighborhood starts to undergo racial transition. If many blacks want to move into the neighborhood, landlords may be able to increase rental income by renting to blacks—indeed, they may be able to increase it enough to compensate them for the costs of extensive turnover. And once racial transition has started, a real estate broker cannot be held accountable and has no reason to discriminate in order to protect his or her reputation.

The second hypothesis, called the agent-prejudice hypothesis, is that landlords' and real estate brokers' own prejudice leads them to discriminate in order to avoid dealing with blacks.

Thus racial prejudice, of both white residents and of housing agents, is the principal cause of racial discrimination in housing. However, the link between prejudice and discrimination is reinforced and magnified by a variety of economic, social, and institutional factors.[2]

First, several real estate institutions reinforce discriminatory behavior. Real estate brokers depend on both informal and formal cooperation with other brokers. In a white community, brokers who refuse to discriminate against blacks may not receive the needed cooperation from other brokers. This cooperation is often formalized with a multiple listing service (MLS), in which all brokers share information about the houses they are trying to sell. An MLS often covers a set of white suburbs and the only brokers with access to it are brokers located in those communities; brokers who serve the black community do not have access. Moreover, access may be controlled by a local real estate board. In the past, suburban brokers who sold to blacks were denied membership in the local board and hence were denied access to the MLS. This type of problem appears to have receded, although good recent evidence is not available.

Second, recent evidence indicates that some lenders still discriminate against blacks in the provision of home mortgages (Schafer & Ladd, 1981). This discrimination by lenders not only adds to the discriminatory barriers that blacks must overcome, but also gives brokers another reason to discriminate—namely, to obtain cooperation from discriminating lenders.

Finally, white residents sometimes greet new black neighbors with hostility or even violence—another form of racial discrimination. Although less common than twenty years ago, incidents of racial violence against incoming black families still occur in many American cities.[3]

In short, the causes and forms of racial discrimination in housing markets are diverse and interconnected. It is no surprise that this discrimination has proven to be so difficult to eliminate.

THE CONSEQUENCES OF RACIAL DISCRIMINATION IN HOUSING

Racial prejudice and racial discrimination have complex effects on the residential structure of an urban area.[4] Discrimination restricts the access of blacks to largely white areas and therefore is a key cause of racial residential segregation. This direct effect is reinforced by an indirect effect on the search behavior of blacks. As shown conceptually by Courant (1978) and empirically by Weisbrod and Vidal (1981), blacks may not even search for housing in white neighborhoods if they expect to encounter discrimination there.

A household's residential location also depends on its income and other socioeconomic characteristics as well as its racial preferences. To some degree, therefore, blacks and whites live in different locations because blacks tend to have lower incomes than whites and because whites do not want to live with blacks. However, many blacks have higher incomes than some whites and, as we will see, most blacks prefer an integrated neighborhood to an all-black one. So income differences and racial preferences are incomplete explanations for extensive racial residential segregation; to some degree, discrimination also is at work.[5]

Prejudice and discrimination also affect the process of neighborhood racial change.[6] Many neighborhoods experience integration in the short run as blacks move into a white area. But this form of integration is rarely stable. Discrimination limits black's housing options and thereby magnifies black demand in neighborhoods into which blacks are allowed to move. Discrimination also preserves options for whites; that is, it ensures that they have all-white neighborhoods to which they can move. As a result, prejudiced whites will not stay in a neighborhood when blacks start to move in. Without discrimination, and therefore with a much wider dispersal of blacks, the pressures that lead to rapid racial transition would be greatly relieved.

Prejudice and discrimination also affect housing prices. If discrimination restricts the neighborhoods into which blacks can move, it boosts the price of housing in largely black areas. Hence blacks must pay more than whites for equivalent housing. This effect is reinforced by search costs. As Courant (1978) shows, the expectation of encountering discrimination may keep blacks from searching in white neighborhoods even if the price is lower there.

Restrictions on housing choice can also lead to systematic differences in housing quality. If blacks do not have as many options as whites, then they will end up in housing of lower quality than whites with the same income and family characteristics. Moreover, discrimination in suburban areas makes it more difficult for blacks to become homeowners, and the ability of blacks to afford higher quality units may be restricted by racial price differentials.

RECENT EVIDENCE ON PREJUDICE AND DISCRIMINATION

RECENT EVIDENCE ON RACIAL PREJUDICE

The principal source of evidence about racial prejudice is social surveys. This survey evidence about white prejudice over the last 40 years is reviewed by Schuman, Steeh, and Bobo (1985). These authors divide the survey questions into three types: questions about principles of equal treatment, questions about the implementation of these principles, and questions about social distance between blacks and whites.

Answers to the first type of question reveal widespread lessening of whites' racial prejudice since 1940. The time period covered by six of the survey questions extends into the 1980s; answers to five of these questions show that prejudice has continued to drop into the 1980s. In 1968, for example, 25% of whites agreed slightly with the statement that "white people have a right to keep blacks out of their neighborhoods"; by 1982, only 15% of the respondents agreed slightly.

On the other hand, answers to the other two types of questions do not reveal a clear lessening of whites' racial prejudice. Attitudes about implementation show little change, on average, over the 1960 to 1983 period, but answers to some questions suggest a lessening of prejudice while answers to some others suggest an increase in prejudice. In 1973, for example, 34% of whites supported a law barring discrimination in housing; by 1983, 45% supported such a law. In contrast, 26% of whites in 1972 agreed that the government "should make every possible effort to improve the social and economic conditions of blacks and other minority groups"; by 1982, only 18% of whites agreed.

Answers to the social distance questions also paint a mixed picture. Social distance questions about residential integration indicate that whites are now less upset than they were in the past about living in a neighborhood with a low level of integration, but may be more upset about a high level of integration. One survey organization asked whites about their neighborhood preferences. In 1972, 23% preferred a mostly white neighborhood, whereas 31% preferred a racially mixed neighborhood or did not care about racial composition. Using a different survey technique and a slightly different wording, the same organization repeated this question in 1981; 34% said mostly white and 42% said mixed. Another survey asked whites whether they would move "if black people came to live in great numbers in your neighborhood." In 1958, 29% said they might move and 20% said they would not move; in 1978, 33% said they might move and 46% said they would not.

My interpretation of these results is that they describe three separate, and somewhat conflicting, trends in racial attitudes. First, belief in racial equality has clearly increased dramatically among most whites in our society. As Schuman et al. point out, however, the fact that people hold this belief does not indicate how they will behave when this belief is in conflict with other beliefs, such as a belief in the individual rights of whites.

The implementation questions ask whether society should devote more resources toward achieving racial equality and therefore to some degree determine the strength of beliefs in racial equality relative to other claims on scarce resources. On average, the answers to these questions do not reveal a lessening of white prejudice. The second conclusion, therefore, is that white people are not much more willing to strive for racial equality than they were ten or fifteen years ago. One important exception for our purposes is in the area of fair housing; whites are less willing to support discrimination against blacks in housing.

The social distance questions reveal a third point: More whites are willing to live in a racially mixed neighborhood than they were ten years ago—but more whites also prefer an all-white neighborhood. These results suggest that more neighborhoods will be receptive to racial integration but that resistance to black entry will continue to be strong in many places.

A more detailed picture of white and black attitudes toward racial integration comes from a survey conducted by Farley, Bianchi, and Colasanto (1979) in Detroit in 1976 and a 1978 national survey by the U.S. Department of Housing and Urban Development (HUD, 1980). The Farley et al. survey found that three-quarters of the whites would be comfortable with one black family in their neighborhood, but that only about one-quarter of whites would be comfortable in a neighborhood with a black majority. To put it another way, 7% of whites would move when the first black family moved in; two-thirds of the whites would move if their neighborhood reached 50% black.

The Farley et al. survey also found that most blacks prefer an integrated neighborhood with about as many blacks as whites; 63% gave this type of neighborhood as their first choice and 20% gave it as their second choice. On the other hand, a significant minority of blacks preferred an all-black neighborhood, 12% as a first choice and 5% as a second choice. Finally, almost all blacks said they would be willing to move into an integrated neighborhood and 37% said they would consider moving into an all-white neighborhood.

A similar picture emerges from the HUD survey, which found that 57% of blacks, 46% of Hispanics, and 16% of whites preferred a neighborhood that was half white and half minority. In addition, 23% of blacks and 18% of Hispanics preferred a neighborhood that was all or mostly minority, whereas 76% of whites preferred a neighborhood that was all or mostly white.

One key conclusion can be drawn from these surveys: Because so many minority households prefer an integrated neighborhood, residential segregation cannot be maintained by prejudice alone.

RECENT EVIDENCE ON RACIAL DISCRIMINATION IN HOUSING

Until recently, the only way for scholars to verify the existence of discrimination in housing was to document the effects of discrimination on housing markets, such as higher prices for equivalent housing in largely black areas than in largely white areas and more centralized residential locations for blacks than for whites with the same socioeconomic characteristics. Several careful studies, based on data from about 1970, implemented this approach and found strong evidence of discrimination.[7] This approach has not yet been widely applied to data from the 1980s. Existing studies, which are reviewed in the fourth section, do not provide a comprehensive picture of the effects of discrimination to compare to the picture from 1970.

Fortunately, however, a new survey technique, called a fair housing audit, allows researchers to observe discriminatory behavior and therefore to obtain direct, easily interpreted measures of the extent and form of discrimination. This section describes the audit technique and presents the results of five recent audit studies, all of which uncover extensive discrimination.

The Fair Housing Audit Technique

In a fair housing audit, an individual from the white majority and an individual from a minority group, who have been carefully matched on their family and economic characteristics, successively visit a landlord or real estate broker to inquire about housing. Both auditors then record on a survey form what they were told and how they were treated by the housing agent. Discrimination exists if the minority auditor is given less complete information or less favorable treatment than his or her majority teammates.

Information provision and treatment, or just treatment for short, by housing agents takes many forms. Each auditor is told about a certain number of housing units in certain locations, invited to inspect some subset of these units, quoted a certain rent or asking price, given certain signals about credit checks and financing, and so on. Audit studies have examined all of these forms of treatment and others. However, most of these studies focus on one kind of treatment, namely, the provision of information about available housing. The most straightforward (and, as we will see, most common) way to deny minorities access to a certain housing unit is not to let them know that it is available.[8] So this focus is in keeping with the principal objective of fair housing legislation, which is to provide equal access to housing.

The treatment received by a home seeker can vary, of course, for many reasons other than discrimination. A carefully designed and managed fair housing audit controls for other factors so that any difference in treatment between a majority auditor and his or her minority teammate must be due to discrimination. An audit controls for auditor characteristics by assigning similar income and family characteristics to teammates, by pairing auditors with similar indelible characteristics such as age and sex, and by giving all auditors the same rigorous training. Moreover, an audit controls for housing market conditions and for housing agency characteristics by sending teammates to visit the same housing agency and ask about the same advertised housing at approximately the same time.[9]

Treatment also can vary for random reasons. By conducting audits for a random sample of advertised housing units, one can employ standard statistical tests to determine whether random factors are at work.[10] The difference in the treatment received by minority and majority auditors is a measure of discrimination.[11]

Results of Recent Audit Studies

Three high-quality studies have examined the level of discrimination in housing availability: Feins, Bratt, and Holister (1981), which is summarized in Feins and Bratt (1983); James, McCummings, and Tynan (1983); and Holshouser (1984).[12] All three studies employ three measures of housing availability: the number of housing units offered to the auditor as serious possibilities, the number of housing units the auditor was invited to inspect, and the number of housing units the auditor actually inspected. The principal results of these studies are presented in Table 2.1.

The 1981 Boston study by Feins et al. found high levels of discrimination in all three measures of housing availability. For example, black auditors were told about 0.64 fewer apartments on average than were their white teammates. Thus housing search is more costly for blacks than for whites; blacks have to visit many more housing agents than do whites to learn about the same number of housing units. The 1983 Boston study by Holshouser, which was carried out in different parts of the city, also found high levels of discrimination against blacks in both the rental and sales markets.

The 1982 Denver study by James et al. also found discrimination, although not in every case. Hispanic owners and black renters encountered high levels of discrimination on all three measures. Hispanic renters faced discrimination in Hispanic sections of Denver and black owners faced discrimination in Anglo suburbs.

Four high-quality studies have examined the probability that minority housing seekers will encounter discrimination in housing availability: Wienk, Reid, Simonson, and Eggers (1979), Hakken (1979), Feins et al. (1981), and Holshouser (1983). The results are presented in Table 2.2.

Two points, which were first raised by Wienk et al., must be made before we can interpret these results. First, one can estimate either the

TABLE 2.1
Estimates of the Level of Discrimination in Housing Availability
(Number of Housing Units)

Sample	*Measure of Housing Availability*[a] *SERPOS*	*INVINS*	*ACTINS*	*Number of Audits*
Boston, 1981[b]				
Black Renters	.641	.764	.537	156
Black Owners	.624	.526	.335	118
Denver, 1982[c]				
Hispanic Renters	.045	.097	.032	62
Hispanic City Areas	.500*	.400	.100	16
Black Renters	.653*	.369*	.267*	70
Hispanic Owners	.427*	.075	.351*	72
Black Owners	.000	.100	.200	49
Anglo Suburban Areas	.800*	.400	.000	21
Boston, 1983[d]				
Black Renters	.570	.570	.380	21
Hispanic Renters	.920	1.170	.250	12
Southeast Asian Renters	1.610	.780	.670	18
Black Owners	.302	.430	.358	38
Hispanic Owners	.446	.599	.205	25

a. SERPOS = units discussed as serious possibilities; INVINS = units invited to inspect; ACTINS = units actually inspected.
b. Feins et al. (1981). All entries significant at the 1% level.
c. James et al. (1983). Overall figures include audits in minority city neighborhoods, Anglo city neighborhoods, and Anglo suburban neighborhoods. Entries marked with an asterisk are listed as statistically significant (5% level) but other entries may also be, see Yinger (1984).
d. Holshouser (1984). Most entries significant at the 1% level; two significant at the 10% level only. Two audits for black renters not included because they are in a different neighborhood.

gross or the net probability that minority auditors encounter discrimination. The gross probability is the percentage of cases in which the minority auditor encounters unfavorable treatment. The net probability is the gross probability minus the percentage of cases in which the majority auditor encountered unfavorable treatment. This distinction is important because any auditor, minority or majority, may encounter less favorable treatment than his or her teammate for random reasons. For example, a white auditor who arrives too close to the housing agent's lunch hour may be treated more abruptly than his or her black teammate! A net measure assumes that the probability of unfavorable treatment for random reasons is the same for blacks and whites. A gross measure assumes that all unfavorable treatment for blacks is due to intentional discrimination, not random factors. Both assumptions are

TABLE 2.2
Estimates of the Probability of Discrimination in Housing Availability
(in percentages)

	Measure of Housing Availability[a]					
Sample	*Advertised Unit*[b]	*SERPOS*	*INVINS*	*ACTINS*	*Original Index*[c]	*Alternate Index*[d]
40 SMSAs, 1977[e]						
Black Renters						
Net	19	24	–	6	27	28
Gross	30	42	–	27	48	49
Black Owners						
Net	10	30	15	10	15	18
Gross	21	54	46	38	39	47
Dallas, 1978[f]						
Dark-skinned Hispanic Renters						
Net	–	37	29	–	43	–
Gross	–	47	36	–	55	–
Light-skinned Hispanic Renters						
Net	–	15	11	–	16	–
Gross	–	32	24	–	39	–
Boston, 1981[g]						
Black Renters						
Net	27	33	38	29	29	43
Gross	37	51	55	46	39	64
Black Owners						
Net	15	30	24	25	24	37
Gross	21	46	43	38	43	61

(continued)

TABLE 2.2 Continued

Sample	Measure of Housing Availability[a]					
	Advertised Unit[b]	*SERPOS*	*INVINS*	*ACTINS*	*Original Index*[c]	*Alternate Index*[d]
Boston, 1983[h]						
Minority Renters						
Net	18	48	39	33	53	71
Gross	23	56	49	38	59	77
Minority Owners						
Net	–6	30	33	22	10	21
Gross	13	44	40	24	38	51

a. SERPOS, INVINS, and ACTINS are defined in the notes to Table 2.1
b. Whether type of unit requested (national and Dallas studies) or advertised unit (Boston studies) was available.
c. Majority favored on at least one item; minority on none.
d. Majority favored on more items than minority.
e. Wienk et al. (1979). All net measures significant at 1% level.
f. Hakken (1979). Net measures significant at 1% level for dark-skinned Hispanics, not significant for light-skinned Hispanics.
g. Feins et al. (1981). All net measures significant at 1% level.
h. Holshouser (1984). "Minority" includes blacks, Hispanics, and (for renters) Southeast Asians. All net measures significant at 1% level, except the two smallest positive entries for owners (significant at 5% level) and the cne negative entry (not significant).

extreme, so the net and gross measures provide a lower and upper bound, respectively, on the probability that a minority will encounter discrimination.[13]

Second, estimates of the probability of discrimination can be made for a single treatment variable, such as the number of units inspected, or for a set of such variables, such as all the variables that have to do with housing availability. The latter approach, which allows one to account for the possibility that discrimination does not take the same form in every audit, has been based on two alternative assumptions. The first is that the majority auditor was favored in the audit if he or she was favored on at least one action and the minority auditor was favored on none. The second is that the auditor favored in the most actions is favored in the set of actions.

We can now interpret Table 2.2. The first column is the probability that the white auditor but not the black auditor will be told that some housing is available. The next three columns indicate the probability that the white will receive more favorable information than the black on the three measures of housing availability defined earlier. And the last two columns present the two indexes defined above for a set of measures of housing availability. Whenever available, both net and gross measures are reported. All studies find high probabilities that blacks will encounter discrimination. Many of the net probabilities exceed 25% and a few of the gross probabilities exceed 50%.

These results apply to a single visit to a housing agent. As Wienk et al. point out, however, housing seekers typically visit more than one agent; the probability that minorities will encounter discrimination during their housing search is the probability that they will encounter discrimination by at least one agent. If the probability of discrimination on a single visit is 15% (below any gross measure in Table 2.2), the probability of encountering at least one act of discrimination in five visits is 56%.[14] If the probability for a single visit is 50%, then the probability in five visits is 97%.

Although these results are strong, they may understate discrimination, which can exist even when blacks and whites are told about the same number of units. For example, both Wienk et al. and the Boston studies found that blacks faced significant discrimination in information about financing. In Wienk et al., the probability that the agent offered to help the auditor obtain financing was 47% for whites but only 38% for blacks. And in Boston in both 1981 and 1983, whites received far more suggestions about financing options than did blacks. The pattern of discrimination facing blacks in Denver is more complex. In the city of

Denver, James et al. found no discrimination in housing availability, but severe discrimination in financing information. In Anglo areas, for example, Anglos received 39 suggestions about creative financing; blacks received none. In the suburbs, the pattern was reversed; as noted earlier, minorities faced high levels of discrimination in housing availability, but the study found little difference between teammates in financing suggestions. Significant discrimination exists in both Denver and its suburbs, but the forms are different.

Existing audit studies also find that blacks are sometimes steered into certain neighborhoods. Using the national audits conducted by Wienk et al., Newberger (1981) calculated the average racial composition, measured by percentage black, of the tracts in which each auditor was shown housing. In 517 cases, the percentage black was higher for the black auditor; in 299 cases is was higher for the white auditor. This probability that this difference could have arisen by chance is less than 0.1% and Newberger concludes that blacks are steered into largely black tracts.

Pearce (1979) conducted 97 audits of real estate brokers in the Detroit area in 1974 and 1975 and obtained detailed geographic information on the houses inspected by the auditors. These audits uncovered extensive racial steering. In the all-white southern, western, and eastern suburbs of Detroit, none of the housing shown to blacks (compared to 60% for whites) was in the community in which the real estate broker's office was located. Furthermore, virtually all the housing shown to blacks was in and around the sole largely black suburb or in areas of Detroit and its northern neighbors that were largely black or near largely black areas.

RECENT EVIDENCE ON THE CAUSES OF DISCRIMINATION

Because fair housing audits directly observe discriminatory behavior by housing agents, they provide an unprecedented opportunity to determine the circumstances under which this discrimination occurs—that is, to test hypotheses about its causes.

The second section stated two principal hypotheses about the causes of racial discrimination by housing agents.

The customer-prejudice hypothesis predicts that housing agents are more likely to discriminate in all-white areas than in largely black areas or in areas undergoing racial transition. Furthermore, it predicts that housing agents are more likely to discriminate against lower-income blacks or blacks with school-aged children, who are more likely to upset their white customers.[15] The agent-prejudice hypothesis predicts that discrimination does not depend on the location of the audit but does reflect variation in racial prejudice among housing agents. Racial prejudice is higher among older people and men are more prejudiced than women (see Schuman et al., 1985). So this hypothesis predicts that older agents (will discriminate more than younger agents and that male agents (will discriminate more than female agents. Moreover, agents could have stronger prejudice against minorities with certain characteristics; for example, they might be particularly averse to dealing with black men.

Simonson and Wienk (1985) test these two hypotheses using the audits collected by Wienk et al. in 40 metropolitan areas in 1977. They regress a measure of the probability of discrimination on a set of variables describing the agent, the auditor, the neighborhood, and the metropolitan area. Two measures of discrimination were used, namely, the probability measures in columns 1 and 4 of Table 2.2 They find support for both hypotheses; that is, both customer and agent prejudice appear to be causes of discrimination by housing agents.[16] As predicted by the customer-prejudice hypothesis, for example, discrimination is lower against higher-income blacks. And as predicted by the agent-prejudice hypothesis, discrimination is higher by older agents and by male agents.

Yinger (1986, December) uses a similar regression methodology to test these two hypotheses using the 1981 Boston audit data collected by Feins et al.[17] The dependent variables are the levels of discrimination in three measures of housing availability (the measures in Table 2.1) in both rental and sales audits. In most cases, discrimination is high (see Table 2.1) but it does not vary with the characteristics of the auditor or of the housing agent. However, discrimination in actually showing houses for sale is stronger against low-income blacks and black families with children. These results support the customer-prejudice hypothesis; brokers keep out the black households that would most upset their white customers. In addition, black females encounter less discrimination in actual inspections than do black males. This result suggests that agent prejudice, and in particular a relatively strong prejudice against black males, is also at work.

Yinger also examines variation in discrimination across locations. He finds high discrimination in all-white areas and low or nonexistent discrimination in areas undergoing racial transition. These results strongly support the customer-prejudice hypothesis. In all-white areas, housing agents discriminate in order to gain favor with their current or potential white customers. Whites do not move into racially changing areas, however, so housing agents in these areas have no white clientele to attract and nothing to gain from discrimination.

Overall, this evidence points to two primary causes of discrimination in housing. Some housing agents in largely white areas discriminate against blacks because they believe it is in their economic interest to cater to the prejudice of their white customers. Other housing agents discriminate because of their own racial prejudice.[18]

RECENT EVIDENCE ON THE CONSEQUENCES OF DISCRIMINATION

The discrimination uncovered by audit studies denies blacks equal access to housing, confines blacks to central locations, and causes racial residential segregation. As explained in the second section, these direct effects may be accompanied by changes in blacks' housing search behavior and higher housing prices in largely black areas than in largely white areas. Several studies provide recent information on these consequences of discrimination. As explained earlier, an examination of these consequences also provides indirect tests for the continued existence of discrimination.

Racial Residential Segregation

The two best-known indexes of segregation are the "dissimilarity" or D index and the "exposure" or E index. The D index indicates the percentage of the black (or white) population that would have to move in order for the percentage black to be the same everywhere. The E index measures the average degree of contact between blacks and whites in residential areas relative to the contact that would occur if blacks and whites were evenly distributed throughout the urban area. Although these two indexes are not directly comparable, both reach a maximum of 100 when blacks and whites are completely separated.

Taeuber (1983) calculated D indexes for 1970 and 1980 for the 28 central cities with black populations greater than 100,000 in 1980. The average value for this index declined from 87 in 1970 to 81 in 1980. This decline was about the same as the decline between 1960 and 1970. Schnare (1985) calculates E indexes for 90 metropolitan areas in 1970 and 1980. The average value for this index was 58 in 1970 and 52 in 1980, so this index also reveals a decline in segregation. This decline occurred in all regions, although it was smallest in the North Central states, which have the highest levels of segregation to begin with. In an earlier paper with a somewhat larger sample, Schnare (1980) found that the E index increased between 1960 and 1970. In all regions except the West, the declines in the 1970s brought the E index approximately back to where it was in 1960. In the West, the decline from 1970 to 1980 (about 12%) was much larger than the increase from 1960 to 1970 (2%).

Overall, the available evidence reveals continuing high levels of racial residential segregation, with modest but widespread declines in this segregation between 1970 and 1980. No study has yet attempted to separate the effects of racial discrimination, prejudice, and income differentials on the level of racial segregation in 1980 or on the change in segregation between 1970 and 1980. Nevertheless, previous studies have shown that discrimination is a key cause of segregation, so the extensive segregation in 1980 provides indirect confirmation of the conclusion that discrimination has not yet been eradicated.

Housing Prices

Several studies, which are reviewed by Yinger (1979), examine housing prices with data from about 1970. The best studies all find that the price of equivalent housing is higher in largely black than in largely white neighborhoods and that within a given neighborhood blacks pay more than whites for equivalent housing.

These studies also discover that one cannot isolate black-white price differentials without carefully controlling for neighborhood quality and without carefully specifying the racial variables. Blacks tend to live in lower-quality neighborhoods, so if one does not control for this factor one may mistakenly conclude that blacks pay less for housing. All studies with good controls for neighborhood quality find higher prices for equivalent housing in largely black than in largely white neighborhoods. In addition, a careful distinction must be made between interneighborhood price differentials and intraneighborhood price differentials.

Apparently, not one study of racial price differentials has been completed with data collected after 1975, adequate neighborhood controls, and a distinction between inter- and intraneighborhood price differentials. The key problem is that the Annual Housing Survey, which is the basis for most housing studies, does not provide adequate neighborhood variables. One study (Merrill, 1977) meets the above methodological requirements but uses 1973-1975 data from the Housing Allowance Experiment in Pittsburgh. This study finds that prices are slightly higher in largely black than in largely white neighborhoods. One other study (Smith and Mieszkowski, described in Mieszkowski, 1979) uses Houston data for 1977 and appears to have adequate neighborhood controls, but does not estimate an intraneighborhood price differential.

This study found slightly higher prices in largely black neighborhoods using one approach and significantly lower prices in largely black neighborhoods using another approach.[19]

In short, we do not yet know whether the large inter- and intraneighborhood racial price differentials that existed in 1970 still exist in the 1980s.[20]

Housing Quality and Homeownership

Another consequence of discrimination is that blacks must select lower-quality housing than whites with similar socioeconomic characteristics. Some recent evidence on this effect is provided by Bianchi, Farley, and Spain (1982). Using data from the 1977 Annual Housing Survey, these authors compare blacks and whites on four measures of housing quality: the percentage of households in housing units that are structurally inadequate, have less than one room per person, are more than 30 years old, and are owner-occupied. For each measure, Bianchi et al. control for the age and education of the household head, family income and type, region, metropolitan status (central city, suburb, nonmetropolitan), and tenure.

By all four measures blacks have lower-quality housing than whites with similar socioeconomic characteristics in similar locations. Blacks are 2.7 percentage points more likely to live in structurally inadequate housing, 5.9 points more likely to live in overcrowded conditions; 4.6 points more likely to live in old housing; and 9.5 points less likely to be homeowners. These gaps are only slightly smaller for recent movers.

Bianchi et al. also calculate the changes in these gaps between 1960 and 1977. For the first two measures, the gap has been cut about in half; the gap in homeownership has been cut by about one-quarter; and the gap in old housing has increased one-quarter.

This evidence indicates that the impact of discrimination on housing quality has lessened somewhat since 1960, but also confirms that discrimination still forces blacks to live in lower-quality housing units than whites.

IMPLICATIONS FOR PUBLIC POLICY

RECENT FEDERAL FAIR HOUSING POLICY

Federal involvement in fair housing enforcement is based largely on Title VIII of the Civil Rights Act of 1968. This title explicitly prohibits racial discrimination in housing and opens three avenues for enforcement. First, it gives the U.S. Department and Urban Development the power to seek conciliation between the parties in a fair housing complaint. However, this power is weak; HUD cannot force people to stay in conciliation proceedings and cannot impose fines or other criminal penalties on housing agents who are found to have discriminated. Furthermore, if a complaint comes from a location that has a state or local fair housing agency with powers based on legislation that is "substantially equivalent" to Title VIII, then HUD must turn the complaint over to that agency.

Second, Title VIII allows the Justice Department to bring suit against housing sellers who have engaged in a "pattern and practice" of discriminatory behavior. These suits may lead to changes in real estate practices but cannot lead to criminal penalties. During the 1970s, the Justice Department initiated only a handful of these pattern-and-practice cases each year; in the 1980s, such cases have virtually disappeared.

Third, individuals can sue a housing seller who has discriminated against them. In the 1980 Havens case, the U.S. Supreme Court ruled that fair housing groups whose efforts to promote integrated neighborhoods have been thwarted by discrimination and a member of a minority group who has encountered discrimination while conducting a fair housing audit both have standing to sue on the basis of Title VIII.

This set of enforcement mechanisms leads to geographic variation in fair housing enforcement efforts. Depending on the location, enforcement is handled by a HUD regional office, by an activist state or local agency with stronger powers than HUD, or by a poorly managed state or local agency that has a substantially equivalent statute but no enthusiasm for fair housing enforcement.

In 1980, HUD implemented the Fair Housing Assistance Program, which is designed to encourage state and local agencies to do more by providing training, technical assistance, and two phases of financial incentives. Phase I funding is based on number of complaints the agency receives. Phase II funding is based on number of cases closed; that is, on the number of cases that come to a successful resolution. All agencies with statutes substantially equivalent to Title VIII are eligible for assistance. By April 1984, 31 state and 50 local agencies had qualified.

A study by Wallace et al. (1985) examines the complaints received by a sample of 15 agencies that obtained FHAP funding. These agencies processed about 2,000 complaints over the three year study period (including 1,200 during FHAP), of which about 400 were closed. In 18% of the cases filed during FHAP, the agency did not proceed because of a lack of jurisdiction, failure by the complainant to pursue his or her complaint, or some other administrative reason. In the rest of the cases, the agency first attempted to obtain a "predetermination" settlement without an extensive investigation. This attempt was successful in 28% of the cases. When such a settlement could not be reached, the agency then determined whether there was "probable cause" for the complaint, that is, whether there was enough evidence to warrant further attempts at conciliation. In 44% of the cases, the agency decided that there was "lack of probable cause" and the complaint was dismissed. At this stage, only about 7% of the cases remained and conciliation was successful in almost all of them. In some jurisdictions, cases with unsuccessful conciliation went on to public hearings or to court.[21]

In short, 35% (predetermination settlements plus conciliation agreements) of the complaints lead to some form of agreement. Slightly less than half of these cases were settled in the complainant's favor; that is, the complainant received some form of relief. The types of relief included an apology or an affirmative action agreement (26% of the cases), the next available unit (14%), a cash settlement (31%), the contested unit (25%), and cash plus the contested unit (4%). In cases with a cash settlement, the median payment was about $450.

The study by Wallace et al. also investigates the effects of FHAP on complaint outcomes. FHAP appears to have improved case processing procedures and to have lowered the number of complaints left unresolved for more than two years. However, the effects of FHAP on the speed with which normal cases are handled and on the type of settlement obtained are not clear. The study did not find any dramatic differences in these outcomes with and without FHAP.

NEW DIRECTIONS FOR FEDERAL POLICY

In my judgment, FHAP has made a useful, but relatively small, contribution to fair housing enforcement. Even with FHAP, the fair housing enforcement mechanism is based largely on complaints and therefore is both unfair and inefficient. It is unfair because discriminators are not punished unless their victims recognize the discrimination and are willing to pursue a long, expensive, emotionally draining conciliation process. It is inefficient because much discrimination goes undetected; as the audit studies make clear, discrimination is often subtle and indeed may not be recognized by the person who encounters it. How does a black or Hispanic person know when he or she is not being told about an available unit? Furthermore, even with FHAP, fair housing enforcement is not supported by criminal penalties and is therefore bound to be feeble. A more effective federal policy must ensure more uniform enforcement, take the enforcement burden off the victims of discrimination, and impose criminal penalties on proven discriminators.

Uniform Coverage

The federal government is responsible for ensuring that all citizens have access to housing, regardless of the enforcement efforts by their state or local fair housing agency. Assisting state and local agencies through FHAP and other programs is a good management strategy in some cases. But the ultimate responsibility for fair housing enforcement lies with the federal government and it must play a more active role in jurisdictions without a substantially equivalent fair housing law and must step in whenever a substantially equivalent law is not adequately enforced.

Furthermore, the federal government should recognize that success in processing fair housing complaints is a limited measure of enforcement success. A better way to determine whether fair housing laws are being adequately enforced would be to determine whether discrimination exists in the housing market for which an agency is responsible. This determination could be made by carrying out fair housing audits for a random sample of available housing units. If these audits uncover a high level of discrimination, then the fair housing agency is not doing its job and active federal intervention should be triggered.

Take the Enforcement Burden Off the Victims

In order to take the enforcement burden off the victims, federal policy needs to move in two directions. First, the federal government should carry out fair housing audits in areas not covered by an effective, substantially equivalent agency and should encourage and support audits by state and local agencies. Audits have been declared legal by the U.S. Supreme Court and are an excellent method for gathering evidence in response to a complaint and thereby improving the conciliation process. Furthermore, audits provide a means for identifying discrimination without waiting for victims to complain. Audits conducted on a random sample of housing units in a housing market generate information on discrimination independent of complaints. And housing agents in this sample who are found to discriminate can be investigated further with more targeted audits. This design, a random sample plus follow-up audits of suspected discriminators, was used in the 1983 Boston study (Holshouser, 1983).

Second, the federal government could take steps to improve the access of all people to information about available housing. The audits studies provide ample evidence that control over this flow is the key tool used by discriminators. The federal government should investigate a variety of methods for improving access, such as required posting, in a central location, of all available housing units and open access for all brokers and fair housing groups to the information in multiple listing services.

Criminal Penalties

The lack of federal criminal penalties for violations of Title VIII is unconscionable. How can we expect to eliminate discrimination in housing when the federal government refuses to punish people who practice it? (A few state and local agencies can impose fines on violators, but the maximum penalty is small.) The federal government should pass a law that allows fair housing enforcement officials to impose heavy fines, at least up to $10,000, on housing sellers who are found, after the conciliation process, to have practiced discrimination. These fines could of course be appealed in court.

Conclusion

Recent studies provide compelling evidence that discrimination is still widely practiced in the housing market. Stronger federal enforcement measures, including widespread use of fair housing audits and criminal penalties for discriminators, are needed to ensure that all citizens have equal access to housing.

NOTES

1. For a review of this literature, see Yinger (1979).

2. One kind of discrimination can lead to another. Because of the high cost of searching for housing, housing sellers have some market power, and a seller with market power will charge a higher price to a group with fewer alternatives. If some sellers refuse to rent or sell to blacks, then blacks have fewer alternatives and this result applies. In other words, discrimination by a few sellers gives all other sellers, even those without prejudice and without prejudiced white customers, an economic incentive to charge blacks a higher price than whites for the same housing; see Yinger (1979).

3. In one 1985 incident, for example, a black women in Cleveland was killed when whites upset about her moving into their neighborhood threw a fire bomb through her window; see National Neighbors (1985).

4. For a more detailed review of these effects, see Yinger (1979) or Yinger et al. (1979).

5. Many studies have tried to distinguish the effects on segregation of prejudice, discrimination, and income. One careful study by Galster (1987) directly measures the powerful effect of discrimination on segregation. Galster also shows that many studies attribute too much of existing segregation to income differences between blacks and whites because they ignore the fact that residential segregation lowers blacks' access to jobs—and hence lowers their incomes.

6. For a more detailed discussion of these issues, see Yinger et al. (1979) or Yinger (1986).

7. See Yinger (1979).

8. As explained in footnote 2, housing sellers may also charge blacks more than whites for equivalent housing. Although there was some evidence for this type of discrimination in 1970 (see Yinger, 1979), none of the audit studies uncover significant discrimination in quoted rents or sales prices.

9. For more on audit study design, see Yinger (1984, forthcoming).

10. Wienk et al. (1979), Feins et al. (1981), and Yinger (1984) examine the issue of sampling.

11. In statistical terms, this difference is an unbiased estimate of discrimination against minorities. A paired difference-of-means test must be used to avoid bias in the standard error of estimated discrimination. A regular difference-of-means test might lead one to conclude that discrimination does not exist when in fact it does; see Yinger (1984, forthcoming).

12. Audits conducted by fair housing groups have uncovered high levels of discrimination in several other cities. For example, Metro Fair Housing Services in Atlanta carried out many audits of one large landlord during 1981 to 1983. I analyzed 71 of these audits. On average, blacks were told about 0.44 fewer available apartments than whites, a difference that is statistically significant at the .05% level. Newspapers have also sponsored and published audit studies in New York, San Jose, Bergen County, New Jersey, Saginaw, Michigan, Washington, D.C., and Miami; see Newberger (1984).

13. When random factors favor the majority auditor, they increase the *severity* of discrimination against minorities but may not alter the *probability* that a minority will encounter discrimination. Yinger (1984) argues, therefore, that a gross measure is a better estimate of the probability that a minority will encounter systematic discrimination than is a net measure.

14. Let P_1 be the probability of encountering discrimination on one visit and P_n be the probability of encountering discrimination at least once in n visits. Then as long as visits are independent, $P_n = 1 - (1 - P1)^n$. See Wienk et al. (1979, p. 126).

15. Racial prejudice is magnified by income-class differences; prejudice in housing is magnified by school integration. See Schuman et al. (1985) and Simpson and Yinger (1985).

16. Simonson and Wienk also examine a third hypothesis, namely, that housing agents discriminate against blacks because they think blacks are less likely to complete a transaction—even controlling for income and family characteristics.

17. The regression techniques of the two studies are similar but not identical; see Yinger (1986, December).

18. These two studies both find support for both hypotheses but differ on a few results. For example, Simonson and Wienk find that the presence of children lowers discrimination; Yinger finds that it increases discrimination. The differences may arise because

Simonson and Wienk examine data for many metropolitan areas, whereas Yinger examines data only for Boston.

19. Both approaches explain house values as a function of housing characteristics. The difference is whether the coefficients of these characteristics are constrained to be the same in both types of neighborhoods; without the constraints, prices are higher in largely black neighborhoods. In my view, unlike the authors', the constraints are not justified and the results without constraints are more compelling.

20. Although recent measures of price differentials are not available, the discrimination uncovered by audits could lead blacks to pay more. Courant (1978) shows that prices may be higher in largely black areas if blacks are denied access to some of the housing in white areas. In the 1981 Boston audits, blacks were denied information about (and hence access to) many housing units. Schafer (1979) found large price differentials, up to 34% for some types of housing, between largely black and largely white areas of Boston in 1970. The audit results and the Courant analysis suggest that these price differentials may have continued into the 1980s.

21. The number and nature of complaints, and of their outcomes, were similar during the pre-FHAP period and in the regions where complaints were processed by HUD; see Wallace et al. (1985).

REFERENCES

Bianchi, S. M., Farley, R., & Spain, D. (1982, February). Racial inequalities in housing: An examination of recent trends. *Demography, 19,* 37-51.

Courant, P. N. (1978, July). Racial prejudice in a search model of the urban housing market. *Journal of Urban Economics, 5,* 329-345.

Farley, R., Bianchi, S., & Colasanto, D. (1979, January). Barriers to the racial integration of neighborhoods: The Detroit case. *Annals of the AAPSS, 441,* 97-113.

Feins, J. D., & Bratt, R. G. (1983, Summer). Barred in Boston: Racial discrimination in housing. *APA Journal,* pp. 344-355.

Feins, J. D., Bratt, R. G., & Hollister, R. (1981). *Final report of a study of racial discrimination in the Boston housing market.* Cambridge, MA: Abt.

Galster, G. C. (1987, January). Residential segregation and interracial economic disparities: A simultaneous equations approach. *Journal of Urban Economics, 21,* 22-44.

Hakken, J. (1979). *Discrimination against Chicanos in the Dallas rental housing market: An experimental extension of the housing market practices survey.* Washington, DC: Department of Housing and Urban Development.

Holshouser, W. (1984). *Final report of a study of racial discrimination in two Boston housing markets.* Cambridge, MA: Abt.

James, F. J., McCummings, B. L., & Tynan, E. A. (1983). *Discrimination, segregation, and minority housing conditions in sunbelt cities: A study of Denver, Houston, and Phoenix.* Denver: University of Colorado at Denver, Center for Public-Private Sector Cooperation, Graduate School of Public Affairs.

Merrill, S. (1977). *Draft report on hedonic indices as a measure of housing quality* (Contract Report H-2040R). Cambridge, MA: Abt.

Mieszkowski, P. (1979). *Studies of prejudice and discrimination in urban housing markets: Special study.* Boston: Federal Reserve Bank of Boston.

National Neighbors. (1985, Summer). Racial violence takes life of Cleveland woman. *National Neighbors Newsletter,* p. 2.

Newberger, H. (1981). *The nature and extent of racial steering practices in U.S. housing markets*. Unpublished manuscript.

Newberger, H. (1984). *Recent evidence on discrimination in housing*. Washington, DC: Department of Housing and Urban Development.

Pearce, D. M. (1979, February). Gatekeepers and homeseekers: Institutional patterns in racial steering. *Social Problems, 3,* 325-342.

Schafer, R. (1979, April). Racial discrimination in the Boston housing market, *Journal of Urban Economics,6*, 176-196.

Schafer, R., & Ladd, H. F. (1981). *Discrimination in mortgage lending*. Cambridge, MA: MIT Press.

Schnare, A. B. (1980, May). Trends in residential segregation by race: 1960-1970. *Journal of Urban Economics, 7*, 293-301.

Schnare, A. B. (1985, October). Trends in Illinois racial segregation in housing: 1960-1980. *Illinois Business Review, 42,* 4-6.

Schuman, H., Steeh, C., & Bobo, L. (1985). *Racial attitudes in America*. Cambridge, MA: Harvard University Press.

Simonson, J., & Wienk, R. E. (1984). *Racial discrimination in housing sales: An empirical test of alternative models of broker behavior*. Unpublished manuscript.

Simpson, G. E., & Yinger, J. M. (1985). *Racial and cultural minorities: An analysis of prejudice and discrimination* (5th ed.). New York: Plenum.

Taeuber, K. (1983). *Racial residential segregation, 28 cities, 1970-1980* (Center for Demography and Ecology Working Paper 83-12). Madison: University of Wisconsin.

U.S. Department of Housing and Urban Development. (1979). *The 1978 HUD survey on the quality of community life: A data book*. Washington, DC: Author.

Wallace, J. E., Holshouser, W. L., Land, T. S., & Williams, J. (1985). *Evaluation of the fair housing assistance program*. Cambridge, MA: Abt.

Weisbrod, G., & Vidal, A. (1981, June). Housing search barriers for low-income renters. *Urban Affairs Quarterly, 16,* 465-482.

Wienk, R. E., Reid, C. E., Simonson, J. C., & Eggers, F. C. (1979). *Measuring discrimination in American housing markets: The housing market practices survey*. Washington, DC: Department of Housing and Urban Development.

Yinger, J. (1979). Prejudice and discrimination in the urban housing market. In P. Mieszkowski & M. Straszheim (Eds.), *Current issues in urban economics*. Baltimore: Johns Hopkins.

Yinger, J. (1984, December). *Measuring racial discrimination with fair housing audits: A review of existing evidence and research methodology*. Paper presented at the HUD Conference on Fair Housing Testing.

Yinger, J. (1986). On the possibility of achieving racial integration with subsidized housing. In J. M. Goering (Ed.), *Housing desegregation, race and federal policy*. Chapel Hill: University of North Carolina Press.

Yinger, J. (1986, December). Caught in the act: Measuring racial discrimination with fair housing audits. *American Economic Review, 76*, 881-893.

Yinger, J., Galster, G. C., Smith, B. A., & Eggers, F. (1979). *The status of research into racial discrimination in American housing markets: A research agenda for the Department of Housing and Urban Development* (Occasional Papers in Housing and Community Affairs, Vol. 6). Washington, DC: Department of Housing and Urban Development.

3

Housing Market Discrimination and Black Suburbanization in the 1980s

JOHN F. KAIN

DURING THE PERIOD 1970-1980, Chicago's suburban black population (SMSA minus the central city) grew by 102,528 or by 80%. Even larger increases were recorded in the Atlanta, Los Angeles, and Washington, D.C. SMSAs and the suburban black population of all United States metropolitan areas grew by almost 1.8 million. These and similar statistics from the 1970 and 1980 censuses have caused a number of observers to conclude that efforts to reduce or eliminate housing market discrimination have at last borne fruit and that large numbers of black Americans are obtaining housing in all-white or predominantly white areas that were previously barred to them.

This chapter uses the Annual Housing Survey (AHS) for 1975 and data from the 1970 and 1980 Censuses of Population and Housing to evaluate the extent and causes of black residential segregation in Chicago and other metropolitan areas and to assess the nature and implications of the recent rapid growth of the suburban black population. In particular, the chapter considers whether the intense segregation of Chicago's black households and those in other United States metropolitan areas can be explained by their lower incomes or other sociodemographic differences between them and nonblack households. Finally, the chapter considers the recent suburbanization of black

---

AUTHOR'S NOTE: *This chapter is a revision of a paper prepared for a Chicago Urban League Conference, "Civil rights in the Eighties: A Thirty Year Perspective," held in Chicago in May 1984.*

households and finds evidence of a new minority residence pattern that offers the first real hope that the intense and destructive pattern of racial residential segregation that has characterized American cities may diminish.

EXTENT OF SEGREGATION

Black Americans have been and remain intensely segregated from whites. Indexes of dissimilarity shown in Table 3.1 for Chicago and all United States metropolitan areas during the period 1940-1970 measure the extend to which observed patterns of residence location by race differ from proportional representation. A value of zero indicates a completely uniform distribution, that is, the proportion of blacks residing in every block or census tract is the same and equal to their proportion of the entire city or metropolitan area population. Conversely, a value of 100 indicates complete segregation of blacks and whites—all blocks or tracts are either 100% or 0% black.

Table 3.1 contains both block and tract indexes. Block indexes are generally regarded as superior measures of the extent of segregation, but since block statistics are unavailable for many parts of metropolitan areas, and particularly before 1970, analysts have computed segregation indexes for both census blocks and tracts. The heavy concentration of black households in central cities and other areas with block statistics also means that exclusive reliance on segregation indexes calculated from block data can be misleading, particularly when trends in the extent of segregation are at issue.

The indexes for Chicago in Table 3.1 demonstrate that Chicago's black population has been intensely segregated since 1940 both absolutely and relative to levels in other metropolitan areas. The indexes for other metropolitan areas are both significantly lower than those for Chicago and exhibit somewhat of a decline between 1960 and 1970. Van Valey, Roof, and Wilcox (1977), however, show that these apparent declines are statistical artifacts that arise from adding smaller metropolitan areas with lower levels of segregation to the 1970 sample.

DETERMINANTS OF SEGREGATION

The most common explanations for the intense segregation of blacks are as follows:

TABLE 3.1
Indexes of Black (Nonwhite) Residential Segregation, Chicago SMSA, 1940-1970

	Year			
Place	*1940*	*1950*	*1960*	*1970*
SMSA (Tract)				
Taeuber		88.1	89.8	
VV-R-W				91.2
Chicago (Tract)				
Taeuber		89.0	89.8	
VV-R-W			91.0	91.8
Chicago (Block)	95.0	92.1	92.6	
Evanston (Block)	91.5	92.1	87.2	
Mean Indexes				
SMSA's (Tracts)				
All			75.4	69.5
North Central			79.5	74.1
Cities (Tract)				
All			75.1	68.3
North Central			75.9	68.5
Cities (Blocks)				
All Cities			86.2	
North Central			87.7	

SOURCES: Van Valey, Roof, & Wilcox (1977) Seggregation indexes computed for 237 fully tracted metropolitan areas in 1960 and 1970. Taeuber & Taeuber (1965).

(1) Socioeconomic differences—black households have lower incomes and differ from white households in other respects that cause them to choose residences in older, more deteriorated neighborhoods. Since low-quality housing tends to be concentrated in the older, more run down portions of central cities, blacks tend to concentrate in these neighborhoods.
(2) Preferences—blacks prefer to live near other blacks, and whites are unwilling to live near blacks. The argument that the intense residential segregation of blacks is attributable to their preferences to live near other blacks is often supported by references to the experience of various nationality groups, who even currently exhibit some tendency to congregate. Proponents of this view also rely heavily on theoretical analyses by Schelling (1969, 1972) that seem to suggest that racially integrated living patterns are inconsistent with current preferences for neighborhoods of varying racial compositions.
(3) Exclusion—the residence choices of blacks are seriously limited by discriminatory practices and the intense segregation of blacks is due in large part to these practices. While individual whites may sometimes refuse to sell or rent to individual blacks, the discriminatory practices of brokers, lenders, and other agents are crucial to the maintenance of

> segregated living patterns. While changes in law and community attitudes have caused these practices to become less blatant, they nonetheless continue to exist.

While each of the preceding explanations partially explains the residence patterns of black Americans, only systemic racial discrimination, that is, exclusion, can explain the intense, unprecedented, and persistent residential segregation of black Americans in Chicago and other United States metropolitan areas. We now turn to these alternative explanations, starting with the most tractable, the so-called socioeconomic hypothesis.

SOCIOECONOMIC DIFFERENCES

A large number of empirical studies have considered whether existing patterns of racial residential segregation can be explained by income and other socioeconomic differences between black and white households. While these studies have consistently shown that the intense segregation of black households cannot be explained by these factors, the myth that income differences are a major, if not the principal, explanation of racial residential segregation persists. Thus we once again examine this proposition using data for the Chicago and Cleveland metropolitan areas.

The Chicago analyses are based on 1975 data from the Spatially Disaggregated Annual Housing Survey (AHS) tape and 1980 data from the Census of Population and Housing. Use of AHS micro (individual household) data permits us to use a sophisticated model that includes household income, family type, family size, and age of head to predict the residence patterns of black households in the Chicago metropolitan area. The principal drawback of the AHS data is that only 17 residence locations are identified—the 12 AHS Analysis Districts within the city of Chicago shown in Figure 3.1 and five areas outside the city. The five suburban areas are the rest of Cook County, Dupage County, Lake County, Kane County, and Will and McHenry Counties.

Because the AHS data provide so little information on the spatial pattern of residence by race in Chicago's suburbs, we complement the AHS analysis with a less elaborate analysis of 1980 census data that uses income to predict the residence patterns of black households. Given sufficient time and money, the more elaborate AHS analysis could be replicated by using 1980 census data using census tracts.[1]

Information on family type, family size, age of head, and household

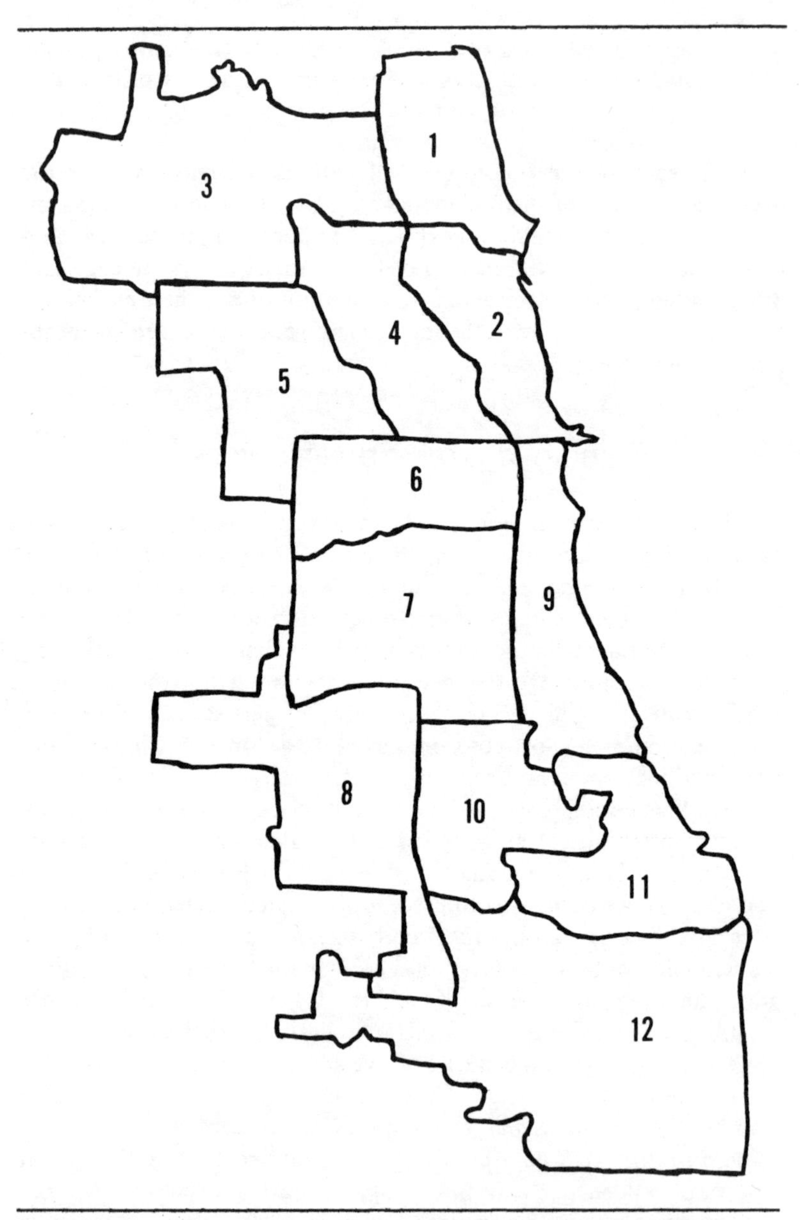

Figure 3.1 Annual Housing Survey Analysis Districts, City of Chicago, 1975

income for individual sample households included in the 1975 Chicago AHS tape was first used to define 216 types of black and white households. Using these household categories, we then computed the fractions black households make up of total households for each household type. As Equations 1 and 2 indicate, the predicted number of black households that would live in each AHS district if race had no affect on residential location decisions, that is, if income were the only determinant of residential location, is obtained by multiplying the SMSA wide proportion black for each household type times the total number of households of each type in cach subarea and finally by summing over the 216 household types.

$$B_{ik} = a_k (H_{ik}) \quad [3.1]$$

$$B_i = Sum_k \, B_{ik} \quad [3.2]$$

where

B = Number of black households

H = Total households

a = Fraction black

i = Analysis districts, i = 1 . . . 15

k = Household types, k = 1 . . . 216.

Table 3.2 displays the results of these calculations for the 12 AHS Chicago districts, the rest of Cook County, and the four County Groups identified on the AHS tapes. Some simple calculations reveal that while nearly 9 out of every 10 Chicago SMSA black households (87.6%) reside in the central city in 1975, only 58.9% would have lived there in the absence of housing market discrimination. Similarly, while 73% of all black households in 1975 lived in the five AHS districts that were more than 65% black in 1975 only 26% would have lived in these areas in the absence of housing market discrimination. Examination of the predicted proportions reveals that none of the Central City AHS districts would have been expected to have fewer than 14.5% black households or more than 40.8% black households in 1975 in the absence of housing market discrimination. In contrast, 3 of the 12 districts had fewer than 4% black households and 5 had more than 81%.

Use of AHS districts significantly understates the actual extent of residential segregation. If the Chicago AHS analyses had employed

TABLE 3.2
Actual and Predicted Numbers of Black Households
and Actual and Predicted Percentage Black by Subarea:
Chicago SMSA in 1975

Subarea	*Number of Black Households* Actual	Predicted	*Percentage Black of All Households* Actual	Predicted
Chicago				
AHS Dist 1	11,328	23,852	9.1	19.1
AHS Dist 2	13,498	24,020	10.8	19.2
AHS Dist 3	1,722	14,346	1.7	14.5
AHS Dist 4	3,797	13,181	3.8	23.0
AHS Dist 5	26,126	20,795	29.6	23.5
AHS Dist 6	50,196	23,315	87.9	40.8
AHS Dist 7	4,664	17,312	6.2	22.9
AHS Dist 8	1,710	13,696	2.0	16.0
AHS Dist 9	68,962	26,407	81.7	31.3
AHS Dist 10	68,854	22,266	96.6	31.3
AHS Dist 11	74,234	23,520	86.7	27.5
AHS Dist 12	45,600	16,661	64.6	23.6
Entire City	370,691	249,371	34.7	22.4
Rest of Cook County	34,856	102,043	4.9	24.3
Dupage County	2,668	22,127	1.6	13.1
Lake County	3,515	13,585	4.0	15.6
Kane County	6,371	17,275	5.5	15.0
Will and McHenry Counties	5,199	18,899	4.3	15.6
Entire SMSA	423,300	423,300	100.0	100.0

SOURCE: Kain, 1980.

census tracts rather than the much larger and more heterogeneous AHS districts, the distribution of tracts by actual and predicted percentage black would be similar to that shown for the Cleveland SMSA in Figure 3.2. This graph summarizes the results of an analysis using the same methodology as described above, except that 1970 Census data and 384 types of households were used. As these data demonstrate, virtually all Cleveland SMSA census tracts in 1970 were either more than 70% or fewer than 4% black, only 313 of the SMSA's 414 tracts were less than 4% black and only 61 were between 5% and 79% black. Most of the 61 "integrated" tracts, moreover, were "transitional" tracts located at the periphery of the ghetto.

In contrast to the actual distributions, 302 of Cleveland's 440 tracts are predicted to have between 10% and 19% black households; only 11 Cleveland tracts had this racial composition in 1970. No Cleveland

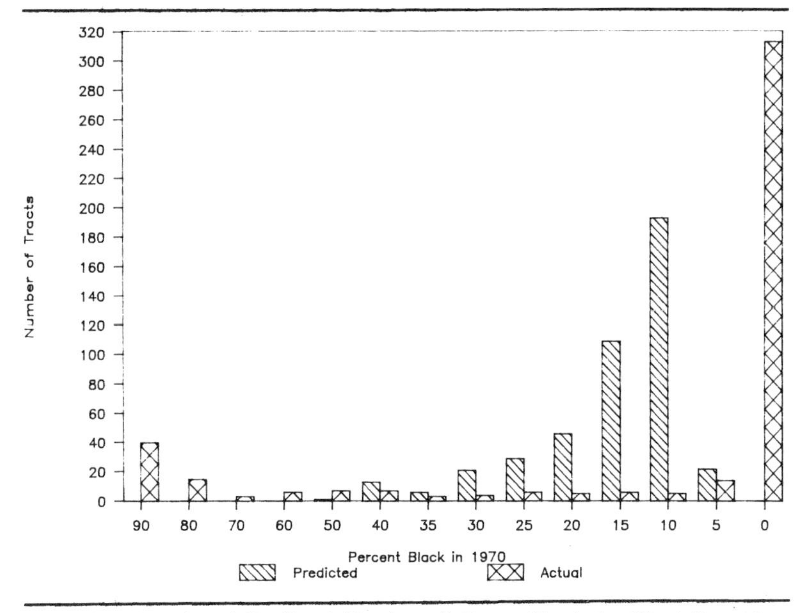

Figure 3.2 Actual and Predicted Number of Census Tracts by Tract Percentage Black, Cleveland SMSA in 1970

tracts had predicted percentages black in excess of 59% as contrasted with 70 actual tracts in this range. As a first approximation, census tracts in Cleveland, in Chicago, and in all other U.S. metropolitan areas are either all black or all white, with small numbers of transitional "integrated" tracts. The predicted proportions, in contrast, cluster tightly around the SMSA mean proportion black with virtually all tracts falling in the range 7%-30% black.

The tendency of the 12 Chicago AHS districts to understate the extent of racial segregation seriously is even more pronounced for suburban Chicago, where the AHS identifies only the rest of Cook County and four other counties or county groups. The Appendix presents the results of an analysis based on 1980 Census data that uses nine income categories to predict the numbers of black households that would reside in the city of Chicago, in each of 122 suburban communities, and in the unincorporated parts of each of the six suburban counties in 1980 in the absence of housing market discrimination. Summaries of the detailed statistics in the Appendix are shown in Table 3.3 for each of 14 categories of suburban communities plus Chicago.

The 14 categories of suburban communities shown in Table 3.3 are

TABLE 3.3
Actual and Predicted Numbers, Percentages, and Cumulative Percentages of Black Households in 1980 by Community Percentage Black: Chicago SMSA, 1980

1980	*Number*	*Percentage Black Households*		*Number Black Households*		*Percentage of*		*Percentage of*		*Cummulative Percentage*	
						SMSA	*Blk Hse*	*Sub*	*Blk Hse*		
% Black	*Places*	*Actual*	*Pred*	*Actual*	*Pred*	*Actual*	*Pred*	*Actual*	*Pred*	*Actual*	*Pred*
Chicago (34.9%)	1	34.9	21.1	381,601	230,823	85.1	51.5				
59.2-70.6	3	64.6	19.0	15,048	4,431	3.4	1.0	22.6	2.0	22.6	22.0
24.6-36.0	3	29.0	18.6	7,381	4,737	1.6	1.1	11.1	2.2	33.6	4.2
14.7-18.2	5	16.8	18.1	14,893	16,047	3.3	3.6	22.3	7.4	56.0	11.6
10.2-11.8	6	10.7	16.2	5,046	7,676	1.1	1.7	7.6	3.5	63.5	15.1
5.1- 8.9	8	6.5	16.2	17,355	43,517	3.9	9.7	26.0	20.0	89.5	35.1
4.0- 4.7	4	4.4	16.8	1,115	4,308	0.2	1.0	1.7	2.0	91.2	37.1
3.1- 3.7	2	3.5	17.7	470	2,405	0.1	0.5	0.7	1.1	91.9	38.2
1.5- 2.3	10	1.9	15.1	1,232	9,786	0.3	2.2	1.8	4.5	93.8	42.7
1.0- 1.4	10	1.3	14.3	1,440	16,418	0.3	3.7	2.2	7.5	95.9	50.3
0.5- 0.9	21	0.7	14.4	2,113	42,985	0.5	9.6	3.2	19.8	99.1	70.0
0.3- 0.4	12	0.3	14.3	351	14,772	0.1	3.3	0.5	6.8	99.6	76.8
0.1- 0.2	25	0.1	14.7	219	24,558	.0	5.5	0.3	11.3	99.9	88.1
0.0- 0.0	13	.0	17.0	36	25,840	.0	5.8	0.1	11.9	100.0	100.0
All Places	123	18.0	18.0	448,300	448,303	100.0	100.0	100.0	100.0	100.0	100.0

defined in terms of the actual racial composition of Chicago's suburban communities in 1980. The second column in Table 3.3 indicates the number of communities in each interval while the third and fourth give the mean actual and predicted percentage black for each category. In spite of somewhat different samples, substantial changes in the region between 1975 and 1980, and other factors, the AHS estimate of the actual black percentage of all Chicago's households in 1975, 34.7%, is quite close to the 1980 Census estimate, 34.9%. The predicted percentages are also quite similar; the AHS analysis obtained a predicted percentage black of 22.4 as contrasted with 21.1 for the 1980 Census analysis.

Further examination of Table 3.3 reveals that only six of Chicago's suburban communities were more than 18.2% black in 1980 (the SMSA-wide percentage was 18.0%) and that only 17 had as many as 10.2% black households. In contrast, 91 were less than 2.4% black and 71 were less than 1% black in 1980. As in the Chicago AHS and Cleveland analyses, the predicted percentages of black households for the 13 categories of suburban communities classified by actual percentage black are highly uniform, ranging from a low of 14.3% for the intervals 1.0%-1.4% and 0.3%-0.4% to a high of 19.0% for the interval 59.2%—70.6%.

The last two columns give the cumulative percentages of suburban black households and illustrate the marked differences between actual and predicted black residence patterns. While the 25 suburban communities with the highest percentage black account for nearly 90% of all suburban black households in 1980, their predicted share was only 35%. It should be understood in making these comparisons, moreover, that the predicted distribution of black suburban residents includes nearly 151,000 more black households than the actual distribution. Columns 7-10 in Table 3.3 show the actual and predicted percentages of black households residing in each category as a fraction of the total black SMSA and total suburban black households, further documenting the marked difference between the actual residence patterns of black households and those that would be expected in the absence of racial discrimination.

PREFERENCES, ATTITUDES, AND EXPECTATIONS

The intense racial segregation of blacks is often justified as a healthy manifestation of a pluralistic society and by reference to the living patterns of various ethnic and nationality groups, who even today exhibit some tendency to cluster within American cities. Nathan Kantrowitz (1971, p. 41), for example, commenting on racial segregation in Boston, argues that "racial segregation is but an extension of ethnic separation, especially since Asian and Latin ethnics show similar patterns in the contemporary city."

Kantrowitz's conclusion concerning the similarity among the experience of blacks and various ethnic groups is very much of a minority position among social scientists who have attempted to compare the residence patterns of blacks and various ethnic minorities. None would disagree with the observation that various ethnic and nationality groups have tended to cluster in identifiable ethnic and nationality group enclaves, but the intense segregation of blacks goes far beyond a tendency to cluster. No ethnic or nationality group currently exhibits levels of segregation as high as those for blacks and, as the indexes of segregation for blacks and various old and new ethnic groups shown in Table 3.4 demonstrate, the segregation of these groups decline rapidly over time, while those for blacks have remained at exceptionally high levels.

Unlike blacks, who as a first approximation live in only all-black or transitional neighborhoods, only a minority of the members of the various identifiable ethnic and nationalities live in ethnic neighborhoods and communities. Identifiable ethnic neighborhoods and communities exist in all American cities, but members of these groups live in all parts of the metropolitan area. As a result, analyses of the residential choices of Cleveland's German, Polish, Czech, Austrian, Hungarian, Yugoslavian, Italian, and Russian-Americans using the methods described previously for Cleveland's blacks are quite successful in predicting the residential distributions of these groups (Kain, 1977). In contrast to Cleveland's blacks, who are highly concentrated in a small fraction of census tracts and communities, members of Cleveland's major nationality groups reside in every community at rates that are very close to their fraction of the region's population. Instead of providing a rationale for the intense racial segregation of blacks, the current and past experience of various ethnic and nationality groups provide strong evidence that the residence patterns of blacks are not due to their preferences to live in all-black neighborhoods, but rather reflect serious and continuing restrictions on their housing opportunities.

Attitudinal surveys also fail to support the preference argument. While there are serious difficulties of interpretation, these surveys provide little support for the view that the current intense pattern of black segregation is the result of black preferences. Thus 74% of black

TABLE 3.4
Mean Indexes of Residential Segregation of Blacks
and Old and New Ethnic Groups, 1910-1950

		Wards		*Tracts*	
City	*Group*	*1910*	*1920*	*1930*	*1940*
	Old	20.6	23.2	26.2	25.4
Boston	New	53.3	45.8	54.6	49.6
	Black	64.1	65.3	77.9	80.1
	Old	28.4	27.9	30.0	28.4
Buffalo	New	60.3	55.2	55.7	48.1
	Black	62.6	71.5	80.5	82.5
	Old	32.6	29.8	27.7	27.8
Chicago	New	52.6	41.1	47.1	41.4
	Black	66.8	75.7	85.2	79.7
	Old	22.4	21.5	28.4	28.7
Cincinnati	New	54.6	46.0	51.6	49.2
	Black	47.3	57.2	72.8	80.6
	Old	24.2	22.6	28.8	27.0
Cleveland	New	52.2	43.1	51.8	45.7
	Black	60.6	70.1	85.0	86.6
	Old	25.6	26.9	28.4	24.3
Columbus	New	52.7	50.6	57.2	46.3
	Black	31.6	43.8	62.8	70.3
	Old	21.6	21.1	29.3	28.4
Philadelphia	New	57.8	47.7	52.9	48.0
	Black	46.0	47.9	63.4	74.0
	Old	23.5	20.9	25.1	24.1
Pittsburgh	New	51.0	47.7	52.7	46.6
	Black	44.1	43.3	61.4	68.5
	Old	21.3	21.8	26.1	27.6
St. Louis	New	57.2	46.5	61.9	43.4
	Black	54.3	62.1	82.1	85.4
	Old	28.4	23.6	33.2	27.2
Syracuse	New	55.6	50.9	59.2	48.9
	Black	64.0	55.2	86.7	85.8

SOURCE: Lieberson, 1963.
NOTE: Community Areas Used for Chicago.

Americans interviewed in a 1969 Newsweek poll responded that they would rather live in a neighborhood that had both whites and Negroes than in a neighborhood with all-Negro families; only 16% chose an all-black neighborhood (Pettigrew, 1973). Moreover, the percentages preferring all-black neighborhoods had declined while the percentage preferring integrated neighborhoods had increased between 1963 and 1969.

Pettigrew (1973), in a review of eleven different surveys conducted between 1958 and 1969, finds that "when presented with a meaningful choice between an all-black neighborhood and a mixed neighborhood, black residents overwhelmingly favored the latter." The same review article also cites survey evidence of increasing racial tolerance among whites. Specifically, Pettigrew compares the responses to identical questions included on seven National Opinion Research Corporation (NORC) polls administered between 1942 and 1968, and a second set of identical questions included in five Gallup polls conducted between 1958 and 1967. In the NORC surveys, the percentage of whites indicating that it would make a difference to them if a Negro—with just as much income and education as the respondent—moved onto their block, declined from 62% in 1942 to 46% in 1956, to 35% in 1965, and to 21% in 1968 (Pettigrew, 1977). Similarly, 48% of whites Gallup interviewed in 1958 stated they definitely or might move "if colored people came to live next door"; nine years later, the percentage had declined to 35%. An even more recent Gallup survey reveals that the fraction of white households who said they would move if a black family lived next door had declined to only 18% in 1978 (American Institute of Public Opinion, 1978).

At first glance these attitudinal data appear highly encouraging to efforts to promote racial residential integration. A more probing examination by Farley, Bianchi, and Colasanto (1978, 1979) of 743 white and 400 black households in Detroit, however, suggests the underlying attitudes of blacks and whites toward integration are considerably more complex. Black respondents were asked to state their preferences for each of five types of neighborhoods defined by the percentage black. Those blacks who expressed an unwillingness to live in an all-black or all-white neighborhood were asked a series of follow-up questions to determine their reasons. The authors also asked white respondents if they would be uncomfortable in each of four types of neighborhoods defined by racial composition and, if so, whether they would move away and their reasons. The authors also attempted to determine the willingness of whites to purchase a home in a racially mixed neighborhood.

Table 3.5 summarizes the findings of Farley et al. concerning the attitudes of blacks and whites toward moving into neighborhoods of various racial compositions. As these data reveal, 82% of blacks selected a 45% black neighborhood as their first or second choice, and only 5% listed an all-white (7%) neighborhood as their first or second choice; the completely black neighborhood was the first or second choice of only 17% of black households. When questioned about their willingness to

TABLE 3.5
Black and White Attitudes Toward Neighborhoods of Varying Racial Composition (in percentages)

Black	*Neigh 1 or 2 Choice*	*Blacks Willing to Move In*	*Whites* Uncomfortable	*Try Move Out*	*Not Move In*
100	17	69	NA	NA	NA
80	33	99	NA	NA	NA
60	NA	NA	72	64	73
45	82	99	NA	NA	NA
33	NA	NA	57	41	73
20	24	95	42	24	50
7	5	38	24	7	27

SOURCE: Farley, Bianchi, & Colasanto, 1979.

move into each type of neighborhood provided they found a nice house they could afford, moreover, fully 38% of black respondents answered they would be willing to be the first black to enter an all-white neighborhood.

Of the large number of blacks (62%) who would not be willing to move into an all-white area, only a few mentioned an explicit desire to live with other blacks, "the majority—about 90%—expressed an opinion that whites in white areas would not welcome them" (Farley et al., 1978, p. 331). About one-sixth of all black respondents stated that "I might get burned out or never wake up," or "They would probably blow my house up." Farley et al. (1978) conclude that "freed of the fear of racial hostility, we believe that most Detroit area blacks would select neighborhoods that are about one-half white and one-half black."

Statistics on the attitudes of whites, summarized in Table 3.5, moreover, indicate that large fractions of whites would feel uncomfortable in a neighborhood with equal numbers of whites and blacks, that if they lived in such a neighborhood they would try to move out, and that they would be unwilling to move into such a neighborhood. Indeed, only 50% of white respondents stated they would be willing to move into a neighborhood with as few as 3 blacks out of 15 households (20% black). In total, 40% of the whites who said they would move away from an integrated neighborhood gave anticipated declines in property values as their reason. From these analyses, Farley et al. conclude, "When we consider the residential preferences of whites in the Detroit area, the prospects for residential integration seem quite slim." These pessimistic conclusions are traceable in part to Schelling's (1969, 1972) influential papers on black and white preferences and their role in creating racially segregated living patterns. The reason is that the distributions of white and black preferences for neighborhoods of varying racial composition obtained by Farely et al. are similar to those Schelling's analyses suggest would produce segregated living patterns.

In interpreting data on black and white preferences, it is important to keep in mind that the attitudes and stated preferences of both blacks and whites are strongly conditioned by existing patterns of segregation and expectations about their continuation. Because the demand for black housing at the periphery of the ghetto has typically been so great, whites usually have been correct in their expectation that the entry of a few blacks would result in a rapid shift to all-black occupancy. If market forces or public policy begin to make these outcomes less certain, both black and white attitudes and behavior could change radically in a very short time. In particular, if blacks begin to appear in all or most neighborhoods—even if in small numbers—the widely accepted notion that the entry of a few black households into a neighborhood or community signals the start of a certain process that must end in that neighborhood or community becoming an all-black slum will become less credible. Further, if all communities come to have some black residents, prejudiced whites will have nowhere to go to find all-white communities. At the same time, the presence of even small numbers of black households in each community provides encouragement, information, and assistance to other black households considering a residential move or actively searching for housing.

EXCLUSION PRACTICES OF RACIAL DISCRIMINATION

While discriminatory practices are probably less prevalent or at least less visible than in the past, a growing body of systematic evidence attests to their continued presence. In HUD's Housing Market Practices Survey pairs of black and white testers—matched in terms of age, general appearance, income, and family size—conducted 3,264 sales and rental audits in 40 metropolitan areas. The results of these audits were used to construct several indexes of discriminatory treatment including the index of housing availability shown in Table 3.6. As these data show, blacks were less favorably treated in 27% of the rental audits and a somewhat smaller fraction of sales audits. In assessing these and similar data, it should be understood that audits cannot identify many kinds of discriminatory practices and particularly acts of discrimination that occur after the initial contact between buyer and broker. Civil rights legislation, the increasing use of audits, threats of legal action, and

TABLE 3.6
Composite Indexes of Housing Availability for Sales and Rental Housing by the Entire U.S., by Region, and by SMSA Size (in percentages)

	No Difference	*White Favored*	*Black Favored*	*Discriminatory Treatment*
Rental Housing				
Entire U.S.	31	48	21	27
Region				
Northeast	32	44	24	20
North Central	34	50	17	33
South	27	52	21	31
West	34	49	17	32
SMSA Size				
Large	30	49	21	28
Small	35	43	22	21
Sales Housing				
Entire U.S.	37	39	24	15
Region				
Northeast	33	39	29	10
North Central	23	55	22	33
South	46	33	22	11
West	34	39	27	12
SMSA Size				
Large	37	40	23	17
Small	36	36	28	7

SOURCE: Wienk et al. 1979, Tables 3, 5, 26, and 28: pp. 58, 66, 124, and 129.

increasingly frequent judgments have caused brokers to be much more careful in their treatment of apparent black home buyers or renters.

While blacks currently can usually expect courteous treatment from brokers, it is much less likely that they will be given equal access to available units. In particular, steering, a practice in which black home buyers are shown housing in all-black or transitional areas while the same agent shows comparable whites units in all-white areas, remains a common practice. In combination with other discriminatory practices, steering maintains and reinforces existing patterns of housing market discrimination. Audits in Boston (Feins et al., 1980) and Cleveland Heights (Saltman, 1978) detected even higher levels of discrimination than were found in the HUD audits and showed steering to be prevalent. Still other analyses, moreover, have documented high levels of discrimination against blacks attempting to obtain financing (Schafer & Ladd,

1981). In combination, with the effects of past discrimination on black ownership, these practices account for the much lower levels of homeownership of blacks, taking into account their lower incomes and other socioeconomic differences and again reinforce existing patterns of residential segregation (Kain & Quigley, 1972, 1975).

BLACK SUBURBANIZATION

While the residential segregation of American blacks remains nearly total and while discrimination continues to limit severely the housing opportunities of black Americans, analysis of the 1980 Census of Population and annual data from other sources, such as school enrollments, suggests the former nearly absolute barriers to black occupancy of many suburban areas may have been breached.

Statistics shown in Table 3.7 document the extent of black suburbanization in the Chicago SMSA in 1970 and 1980. These data indicate that just over 30,000 black households resided in the 122 suburban areas listed in Appendix A in 1970 and that the number had more than doubled to nearly 67,000 by 1980. Further inspection of the summary statistics in Table 3.7 and of detailed data in the appendix, however, demonstrates that Chicago's black suburban residents remain highly concentrated. As the last three columns in Table 3.7 reveal, the 9 suburban communities with at least 775 black households in 1980 accounted for 68.4% of all suburban black households in 1980. The last row gives the numbers of black households residing in unincorporated portions of the six suburban counties; 79% of the 6,622 black households in 1970 and 64% of the 12,351 in 1980 lived in unincorporated parts of Cook County—most of these areas were extensions of the central city ghetto.

Table 3.8 lists the 42 suburban communities with more than 100 black households in 1980 by their numbers of black households in that year. In total, 9 suburban communities including 5 in Cook County, 2 in Lake County, and 1 in each of Will and Kane County had more than 1,000 black households in both 1970 and 1980; the growth in black population in these 9 established black communities accounted for more than half of the 1970-1980 growth in black households in Chicago's suburbs. Most of the remaining black suburban growth occurred in suburban communities on the periphery of the central city ghetto and represented a continued peripheral growth of the ghetto across the boundaries of the central city.

TABLE 3.7
Actual and Predicted Numbers of Black Households in 1970 and 1980
by Actual Community Percentage Black in 1980

Number Black Households	*Number of Communities*		*Number Black Households*			*Percentage Black Households*			*Cum % All Places*			*Cum % Suburbs*		
			Actual		*Pred*	*Actual*		*Pred*	*Actual*		*Pred*	*Actual*		*Pred*
	1970	*1980*	*1970*	*1980*	*1980*	*1970*	*1980*	*1980*	*1970*	*1980*	*1980*	*1970*	*1980*	*1980*
Chicago	1	1	314,640	381,601	230,823	27.7	34.9	21.1	91.2	85.1	51.5			
5,000-7,977	0	3	0	17,535	8,425	0.0	37.1	17.8	91.2	89.0	53.4	0.0	26.3	3.9
2,521-4,615	3	5	9,540	16,503	14,011	20.4	22.2	18.8	94.0	92.7	56.5	31.5	51.0	10.3
1,753-2,454	4	3	8,229	6,724	9,904	13.3	11.9	17.5	96.4	94.2	58.7	58.6	61.1	14.9
775-1,465	2	5	2,742	4,850	10,959	9.3	7.5	16.9	97.2	95.3	61.1	67.7	68.4	19.9
356-610	3	6	1,700	2,645	4,462	6.4	9.8	16.6	97.7	95.9	62.1	73.3	72.4	22.0
252-339	0	6	0	1,825	8,957	0.0	3.1	15.2	97.7	96.3	64.1	73.3	75.1	26.1
114-246	5	14	802	2,102	20,935	2.4	1.5	14.7	97.9	96.8	68.8	76.0	78.2	35.7
52-99	3	14	198	1,151	18,280	1.3	0.9	14.4	97.9	97.0	72.9	76.6	80.0	44.1
26-49	4	19	144	724	19,837	0.4	0.5	14.3	98.0	97.2	77.3	77.1	81.0	53.2
15-21	4	6	75	104	5,236	0.2	0.3	14.1	98.0	97.2	78.5	77.3	81.2	55.6
10-13	4	5	50	57	4,315	0.1	0.2	15.7	98.0	97.2	79.4	77.5	81.3	57.6
5-9	15	15	103	103	18,855	0.1	0.1	15.6	98.1	97.2	83.6	77.8	81.4	56.3
1-4	46	11	93	25	13,326	.0	.0	16.8	98.1	97.2	86.6	78.1	81.5	72.4
0-0	23	4	0	0	4,040	0.0	0.0	16.4	98.1	97.2	87.5	78.1	81.5	74.3
Remainder	6	6	6,622	12,351	55,940	2.4	3.4	15.3	100.0	100.0	100.0	100.0	100.0	100.0
Total	123	123	344,938	448,303	448,303	15.8	18.0	18.0	100.0	100.0	100.0	100.0	100.0	100.0
	123		344,938	448,300	448,303	15.8	18.0	18.0						

TABLE 3.8
Total and Black Households in 1970 and 1980
and Predicted Numbers of Black Households
in 1980 by Area: Chicago SMSA

			Number Black Households			*Percentage Black Households*		
		1980	*Actual*		*Pred*	*Actual*		*Pred*
Community	*Co.*	*Median*	*1970*	*1980*	*1980*	*1970*	*1980*	*1980*
Harvey	C	18,033	2,793	6,499	2,207	26.0	59.2	20.1
Maywood	C	19,199	3,159	5,970	1,565	36.0	71.4	18.8
Evanston	C	21,715	3,588	5,066	4,653	13.2	18.2	16.7
Joliet	W	18,966	2,361	4,615	5,219	9.3	16.9	19.2
Waukegan	L	19,091	2,276	3,839	4,462	10.9	15.9	18.5
Chicago Hgts.	C	18,585	1,859	2,949	2,314	15.7	24.6	19.3
Markham	C	23,265	1,733	2,579	658	47.1	64.9	16.6
North Chicago	L	15,853	1,465	2,521	1,358	23.6	36.0	19.4
Aurora	K	19,964	1,277	2,454	4,940	5.5	8.9	17.9
Oak Park	C	20,601	30	2,359	3,899	0.1	10.5	17.3
Bellwood	C	22,963	34	1,911	1,065	0.5	29.6	16.5
Elgin	K	19,077	610	1,196	4,217	3.4	5.1	18.0
Calumet	C	21,654	7	1,127	2,674	0.1	7.2	17.1
Park Forest	C	23,553	196	924	1,452	2.3	10.3	16.2
Zion	L	20,593	511	848	1,017	10.5	14.9	17.9
Bolingbrook	W	26,403	0	755	1,599	0.0	6.9	14.6
Summit	C	18,007	579	525	696	15.8	14.7	19.5
Country Club Hills	C	27,239	2	497	577	0.1	11.8	13.7
Justice	C	21,228	3	456	643	0.1	11.7	16.6
Hazel Crest	C	26,272	1	454	643	.0	10.2	14.5
Forest Park	C	17,995	1	357	1,439	.0	4.7	19.0
Matteson	C	27,667	1	356	463	0.1	11.1	14.5
Le Grange	C	27,609	120	338	822	2.4	6.1	14.9
Wheaton	D	27,996	165	334	2,017	1.9	2.3	14.0
Woodridge	D	26,248	7	317	1,144	0.3	4.1	14.6
Blue Island	C	16,664	199	311	1,677	2.4	3.7	19.7
Glenwood	C	28,415	0	273	464	0.0	8.1	13.7
Schaumburg	C	26,270	3	252	2,834	0.1	1.3	14.6
Carol Stream	D	21,353	14	246	1,012	1.2	4.2	17.4
Batavia	K	23,133	122	195	712	4.4	4.5	16.3
Hoffman Estates	C	27,685	5	175	1,699	0.1	1.4	13.9
Lisle	D	26,123	2	159	728	0.1	3.1	14.3
Skokie	C	27,402	15	157	3,238	0.1	0.7	14.5
Mount Prospect	C	27,093	3	151	2,728	.0	0.8	14.6
Highland Park	L	38,542	89	144	1,233	1.0	1.4	12.1
Lansing	C	24,139	4	134	1,638	0.1	1.3	15.8
Dalton	C	25,506	8	133	1,274	0.1	1.6	15.4
Westmont	D	22,824	2	131	1,076	0.1	2.0	16.1
Glendale Heights	D	25,581	7	128	1,090	0.3	1.7	14.7
Hanover Park	C	26,033	0	119	1,254	0.0	1.4	14.4
Downers Grove	D	27,923	21	116	2,207	0.2	0.8	14.5
Homewood	C	27,305	4	114	1,045	0.1	1.6	14.6

While nearly all of Chicago's black suburbanization during the period 1970-1980 was accounted for by the growth of established black communities or peripheral expansion of the central city ghetto, it is also evident that a new and highly encouraging pattern of black residence choices emerged during the decade. Specifically, the number of suburban communities with no black households declined from 23 in 1970 to 4 in 1980 and the number with between 1 and 4 black households declined from 46 to 11. At the same time, the number of Chicago suburban communities with between 26 and 246 black households increased from 12 in 1970 to 47 in 1980 and the number with 252 to 610 black households increased from 3 in 1970 to 12. Even though the 59 black suburban communities with between 26 and 610 black households in 1980 accounted for less than 2% of all black households in the SMSA in 1980, the significance of the appearance of even token numbers of black households in so many, widely dispersed suburban communities cannot be overemphasized.

The difference between no black households and some is enormous. In addition, each of the nearly 8,500 black households living in these communities in 1980 has dozens of black relatives, friends, and coworkers. Their success in overcoming continuing discriminatory barriers is a source of information and encouragement for thousands or possibly hundreds of thousands of other black families, who even now may be considering a move to the suburbs. Further, the appearance of nontrivial numbers of black households in a large number of widely dispersed suburban communities, and, even more important, of growing numbers of black children in previously all-white schools, must have begun to change the expectations of both blacks and whites about the likely impacts of black entry into formerly all-white neighborhoods. While black entry into all-white communities adjacent to the central city ghetto may still signal a nearly certain transition to all or nearly all-black occupancy, there are simply too few blacks to produce this outcome in the large number of widely dispersed communities that had between 26 and 600 black households by 1980.

As both blacks and whites cease to believe that black entry necessarily leads to the creation of an all-black slum, it will occur less often. Whites will continue to buy or rent housing in neighborhoods following black entry and blacks will no longer feel they can obtain improved housing only in a few communities and neighborhoods that have come to be recognized as acceptable for black occupancy. The resulting changes in black and white expectations could produce significant decreases in racial residential segregation with surprising speed.

APPENDIX

Total and Black Households in 1970 and 1980 and Actual and Predicted Numbers of Black Households in 1980 by Area: Chicago SMSA

Areas	1980 Median	*Total Households* 1970	*Total Households* 1980	*Number Black Households* Actual 1970	Actual 1980	Pred 1980	*Percentage Black Hse* Actual 1970	Actual 1980	Pred 1980
Entire SMSA	20,726	2,183,646	2,486,724	344,965	448,303	448,303	15.8	18.0	18.1
Cook County	19,187	1,766,035	1,879,117	334,468	425,127	354,480	18.9	22.6	18.9
Du Page County	27,509	136,251	222,014	355	2,631	31,557	0.3	1.2	14.2
Kane County	22,102	74,642	93,729	2,084	3,933	15,684	2.8	4.2	16.8
Lake County	25,210	102,947	139,715	4,458	7,773	21,667	4.3	5.6	15.5
McHenry County	23,473	33,083	49,078	6	27	7,932	.0	0.1	16.2
Will County	23,329	70,688	103,071	3,594	8,812	16,983	5.1	8.5	16.5
Chicago	15,301	1,137,854	1,093,407	314,640	381,601	230,823	27.7	34.9	21.1
Rest of Cook County	24,143	115,134	122,414	5,205	7,977	19,366	4.5	6.5	15.8
Harvey	18,033	10,751	10,978	2,793	6,499	2,207	26.0	59.2	20.1
Maywood	19,199	8,774	8,357	3,159	5,970	1,565	36.0	71.4	18.8
Evanston	21,715	27,173	27,907	3,588	5,066	4,653	13.2	18.2	16.7
Joliet	18,966	25,342	27,272	2,361	4,615	5,219	9.3	16.9	19.2
Waukegan	19,091	20,965	24,134	2,276	3,839	4,462	10.9	15.9	18.5
Rest of Will Co.	24,420	41,871	59,858	1,225	3,235	9,435	2.9	5.4	15.8
Chicago Hgts.	18,585	11,856	11,980	1,859	2,949	2,314	15.7	24.6	19.3
Markham	23,265	3,683	3,971	1,733	2,579	658	47.1	64.9	16.6
North Chicago	15,853	6,219	6,999	1,465	2,521	1,358	23.6	36.0	19.4
Aurora (pt)	19,964	23,334	27,668	1,277	2,454	4,940	5.5	8.9	17.9
Oak Park	20,601	22,620	22,511	30	2,359	3,899	0.1	10.5	17.3
Bellwood	22,963	6,519	6,463	34	1,911	1,065	0.5	29.6	16.5

Elgin (pt)	19,077	17,881	23,463	610	1,196	4,217	3.4	5.1	18.0
Calumet	21,654	10,505	15,655	7	1,127	2,674	0.1	7.2	17.1
Park Forest (pt)	23,553	8,440	8,985	196	924	1,452	2.3	10.3	16.2
Zion	20,593	4,878	5,696	511	848	1,017	10.5	14.9	17.9
Bolingbrook (pt)	26,403	1,666	10,969	0	755	1,599	0.0	6.9	14.6
Summit	18,007	3,670	3,568	579	525	969	15.8	14.7	19.5
Country Club Hills	27,239	1,499	4,204	2	497	577	0.1	11.8	13.7
Rest of DuPage Co.	29,684	35,018	54,062	63	466	7,279	0.2	0.9	13.5
Justice	21,228	2,989	3,884	3	456	643	0.1	11.7	16.6
Hazel Crest	26,272	2,725	4,454	1	454	643	.0	10.2	14.5
Forest Park	17,995	6,290	7,569	1	357	1,439	.0	4.7	19.0
Matteson	27,667	1,334	3,203	1	356	463	0.1	11.1	14.5
Rest of Kane Co.	25,698	20,845	29,659	70	339	4,441	0.3	1.1	15.0
Le Grange	27,609	5,098	5,507	120	338	822	2.4	6.1	14.9
Wheaton	27,996	8,557	14,379	165	334	2,017	1.9	2.3	14.0
Rest of Lake Co.	25,793	40,692	65,386	56	319	9,918	0.1	0.5	15.2
Woodridge	26,248	2,531	7,823	7	317	1,144	0.3	4.1	14.6
Blue Island	16,664	8,186	8,506	199	311	1,677	2.4	3.7	19.7
Glenwood	28,415	1,917	3,378	0	273	464	0.0	8.1	13.7
Schaumburg (pt)	26,270	4,804	19,488	3	252	2,834	0.1	1.3	14.6
Carol Stream	21,353	1,171	5,835	14	246	1,012	1.2	4.2	17.4
Batavia (pt)	23,133	2,781	4,380	122	195	712	4.4	4.5	16.3
Hoffman Estates	27,685	5,337	12,218	5	175	1,699	0.1	1.4	13.9
Lisle	26,123	1,470	5,090	2	159	728	0.1	3.1	14.3
Skokie	27,402	20,926	22,314	15	157	3,238	0.1	0.7	14.5
Mount Prospect	27,093	9,415	18,769	3	151	2,728	.0	0.8	14.6
Highland Park	38,542	9,186	10,223	89	144	1,233	1.0	1.4	12.1
Lansing	24,139	7,568	10,371	4	134	1,638	0.1	1.3	15.8
Dalton	25,506	6,919	8,287	8	133	1,274	0.1	1.6	15.4
Westmont	22,824	2,907	6,703	2	131	1,076	0.1	2.0	16.1

(continued)

APPENDIX Continued

Areas	1980 Median	Total Households 1970	Total Households 1980	Number Black Households Actual 1970	Number Black Households Actual 1980	Number Black Households Pred 1980	Percentage Black Hse Actual 1970	Percentage Black Hse Actual 1980	Percentage Black Hse Pred 1980
Glendale Heights	25,581	2,607	7,411	7	128	1,090	0.3	1.7	14.7
Hanover Park (pt)	26,033	2,738	8,726	0	119	1,254	0.0	1.4	14.4
Downers Grove	27,923	10,091	15,282	21	116	2,207	0.2	0.8	14.5
Homewood	27,305	5,751	7,149	4	114	1,045	0.1	1.6	14.6
Palos Hills	26,455	1,886	5,607	0	99	814	0.0	1.8	14.5
Lombard	25,911	9,853	12,981	2	95	1,917	.0	0.7	14.8
Rolling Meadows	25,045	5,321	6,905	8	94	1,012	0.2	1.4	14.7
Prospect Heights	22,841	3,529	4,679	3	90	735	0.1	1.9	15.7
Crestwood	22,324	1,280	3,796	57	89	619	4.5	2.3	16.3
Glen Ellyn	29,420	6,360	8,444	18	86	1,201	0.3	1.0	14.2
Addison	24,957	6,658	9,655	6	86	1,479	0.1	0.9	15.3
Wheeling (pt)	23,506	4,014	9,038	1	85	1,419	.0	0.9	15.7
Elk Grove (pt)	28,210	6,104	9,363	5	79	1,299	0.1	0.8	13.9
Glenview	33,078	7,180	10.670	2	77	1,335	.0	0.7	12.5
West Chicago	20,466	2,874	4,099	5	72	685	0.2	1.8	16.7
Arlington Hgts (pt)	30,205	17,804	22,180	5	71	3,013		0.3	13.6
Naperville (pt)	34,141	6,382	13,043	8	70	1,629	0.1	0.5	12.5
Carpentersville	23,075	5,774	6,863	2	58	1,122	.0	0.8	16.4
Lake Forest	44,767	4,216	4,821	52	49	565	1.2	1.0	11.7
Bloomingdale	28,891	741	3,977	0	48	541	0.0	1.2	13.6
Roselle (pt)	28,359	1,151	5,792	0	48	830	0.0	0.8	14.4
Buffalo Grove (pt)	28,219	2,826	7,001	1	48	918	.0	0.7	13.1
Streamwood	26,247	4,185	6,445	0	45	896	0.0	0.7	13.9
Villa Park	24,962	7,239	7,900	3	43	1,230	.0	0.5	15.6
Sauk Village	24,827	1,570	2,847	2	42	431	0.1	1.5	15.2

Alsip	23,212	3,026	6,036	8	40	966	0.3	0.7	16.0
Elmhurst	27,613	14,462	14,722	11	39	2,077	0.1	0.3	14.1
Palatine	30,396	6,688	10,628	3	38	1,450	.0	0.4	13.7
Tinley Park (pt)	25,031	3,116	8,314	0	36	1,293	0.0	0.4	15.6
Des Plaines	25,470	16,790	18,779	13	36	2,878	0.1	0.2	15.3
Bartlett (pt)	29,417	1,030	4,217	2	35	592	0.2	0.8	14.1
Romeoville	25,226	2,784	3,838	0	34	554	0.0	0.9	14.5
Schiller Park	21,787	4,176	4,244	8	32	708	0.2	0.8	16.7
Hinsdale (pt)	35,015	4,846	5,748	21	29	719	0.4	0.5	12.5
South Holland	29,214	6,098	7,506	2	29	1,014	.0	0.4	13.5
St. Charles (pt)	24,493	4,027	6,165	3	27	965	0.1	0.4	15.7
Wilmette	36,980	9,273	9,725	12	26	1,209	0.1	0.3	12.5
Bennsenville (pt)	23,102	3,564	5,853	0	21	923	0.0	0.4	15.8
Darien	31,415	2,106	4,475	0	20	572	0.0	0.4	12.8
Riverdale	20,028	5,556	5,606	4	16	992	0.1	0.3	17.7
Oak Forest	26,289	4,726	7,708	0	16	1,123	0.0	0.2	14.6
Northbrook	39,926	6,876	9,552	2	16	1,069	.0	0.2	11.2
River Forest	30,561	3,848	4,050	1	15	557	.0	0.4	13.8
Rest of McHenry Co.	23,958	23,453	34,538	3	15	5,500	.0	.0	15.9
Northlake	22,233	4,088	4,227	1	13	707	.0	0.3	16.7
Deerfield (pt)	39,076	4,988	5,360	4	13	605	0.1	0.2	11.3
Melrose Park	18,903	7,463	7,982	40	11	1,498	0.5	0.1	18.8
Midlothian	22,283	4,131	4,554	40	10	766	1.0	0.2	16.8
Libertyville	31,815	3,371	5,307	3	10	739	0.1	0.2	13.9
Orland Park	29,558	1,553	6,963	0	9	927	0.0	0.1	13.3
Morton Grove	29,952	7,068	7,949	1	9	927	0.0	0.1	13.3
Franklin Park	22,888	6,180	6,126	2	8	1,013	.0	0.1	16.6
Niles	25,244	8,960	10,232	4	8	1,589	.0	0.1	15.5
Round Lake Beach	22,928	1,573	3,612	0	7	600	0.0	0.2	16.6
Winnetka	48,872	4,152	4,260	9	7	460	0.2	0.2	10.8

(continued)

APPENDIX Continued

Areas	*1980 Median*	*Total Households 1970*	*Total Households 1980*	*Number Black Households Actual 1970*	*Number Black Households Actual 1980*	*Number Black Households Pred 1980*	*Percentage Black Hse Actual 1970*	*Percentage Black Hse Actual 1980*	*Percentage Black Hse Pred 1980*
Chicago Ridge	20,039	2,393	4,973	0	7	879	0.0	0.1	17.7
Park Ridge	31,056	12,898	13,275	2	7	1,791	.0	0.1	13.5
Oak Lawn	24,202	16,522	20,725	1	7	3,290	.0	.0	15.9
Western Springs	34,122	3,549	4,281	3	6	514	0.1	0.1	12.0
Woodstock	18,556	3,349	4,420	3	6	807	0.1	0.1	18.3
Le Grange Park	25,418	5,010	5,171	1	6	792	.0	0.1	15.3
Mundelein	24,610	4,033	5,519	1	6	850	.0	0.1	15.4
Wood Dale	27,473	2,410	3,656	0	5	509	0.0	0.1	13.9
Berwyn	17,992	19,971	19,831	1	5	3,762	.0	.0	19.0
Westchester	27,843	5,849	6,170	1	4	861	.0	0.1	14.0
Crystal Lake	26,021	4,143	6,191	0	4	956	0.0	0.1	15.5
Brookfield	22,992	6,545	7,299	1	4	1,195	.0	0.1	16.4
Cicero	16,730	24,663	24,212	1	3	4,786	.0	.0	19.8
Palos Heights	34,462	2,449	3,242	7	2	398	0.3	0.1	12.3
McHenry	21,601	2,138	3,929	0	2	667	0.0	0.1	17.0
Norridge	24,325	4,788	5,668	0	2	878	0.0	.0	15.5
Lincolnwood	36,463	3,793	4,094	1	1	482	.0	.0	11.8
Hickory Hills	25,380	3,615	4,468	0	1	685	0.0	.0	15.4
Bridgview	22,041	3,343	4,658	1	1	791	.0	.0	17.0
Elmwood Park	21,335	8,883	9,429	1	1	1,626	.0	.0	17.3
Burbank	25,054	0	8,562	0	0	1,328	ERR	0.0	15.5
Worth	21,778	3,490	4,230	0	0	728	0.0	0.0	17.2
Evergreen Park	23,501	7,372	7,567	1	0	1,226	.0	0.0	16.2
River Grove	19,599	3,896	4,254	0	0	758	0.0	0.0	17.8
Sum (est) SMSA	20,726	2,182,508	2,486,721	344,938	448,300	448,303	15.8	18.0	18.1
Actual SMSA	20,726	2,183,646	2,486,724	344,965	448,303	448,303	15.8	18.1	18.1

NOTE

1. In fact, I have performed such an analysis for Kansas City. The overall results are virtually identical to those for Chicago and Cleveland described in this chapter.

REFERENCES

American Institute of Public Opinion. (1978, November), *The Gallup opinion index*. Princeton, NJ: Author.

Farley, R., Bianchi, S., & Colasanto, D. (1978). Chocolate city, vanilla suburbs: Will the trend toward racially separate communities continue? *Social Science Research, 7*, 319-344.

Farley, R., Bianchi, S., & Colasanto, D. (1979, January). Barriers to the racial integration of neighborhoods in the Detroit case. *Annals of the American Academy of Political and Social Science, 441*, 97-118.

Feins, J. D., Bratt, R. C., & Hollister, R. (1982). *Final report of a study of racial discrimination in the Boston housing market*. Cambridge, MA: Abt.

Grier, G., & Grier, E. (1983). *Black suburbanization in the 1970s: An analysis of census results*. Bethesda, MD: Grier Partnership.

Hersberg, T. et al. (1971, January). A tale of three cities: Blacks and immigrants in Philadelphia: 1850-1880, 1930, and 1970. *Annals of the American Academy of Political and Social Science, 441*.

Kain, J. F. (1977). Race ethnicity and residential location. In R. E. Grieson (Ed.), *Public and urban economics*. Lexington: D. C. Heath.

Kain, J. F. (1980). *National urban policy paper on the impacts of housing market discrimination and segregation on the welfare of minorities*. Paper prepared for the Assistant Secretary for Community Planning and Development, U.S. Department of Housing and Urban Development, Cambridge, MA.

Kain, J. F., & Quigley, J. M. (1972, June). Housing market discrimination, homeownership, and savings behavior. *American Economic Review*.

Kain, J. F., & Quigley, J. M. (1975). *Housing markets and racial discrimination: Microeconomic analysis*. New York: National Bureau of Economic Research.

Kantrowitz, N. (1971, January). Racial and ethnic residential segregation in Boston 1830-1970. *Annals of the American Academy of Political and Social Science, 441*.

Lake, R. W. (1981). *The new suburbanites: Race and housing in the suburbs*. New Brunswick: Center for Urban Policy Research.

Lieberson, S. (1963). *Ethnic patterns in American cities*. New York: Free Press.

Pettigrew, T. (1973). Attitudes on race and housing: A socio-psychological view. In A. M. Hawley & V. P. Rock (Eds.), *Segregation in residential areas*. Washington, DC: National Academy of Sciences.

Schafer, R., & Ladd, H. F. (1981). *Discrimination in mortgage lending*. Cambridge: MIT Press.

Saltman, J. (1978, February). *Cleveland Heights: Housing availability survey* (Final Report). Cleveland Heights Community Congress (mimeo).

Schelling, T. (1969, May). Modes of segregation. *American Economic Review*, pp. 169-85.

Schelling, T. (1972). A process of residential segregation: Neighborhood tipping. In A. H. Pascal (Ed.), *Racial discrimination in economic life*. Lexington: Lexington Books.

Schnare, A. (1977). *Residential segregation by race in U.S. metropolitan areas: An analysis across cities and over time* (Urban Institute Working Paper 246-2). Washington, DC.

Taeuber, K., & Taeuber, A. (1964, January). The negro as an immigrant group. *American Journal of Sociology, LXIV*, 374-382.

Taeuber, K., & Taeuber, A. (1965). *Negroes in cities: Residential segregation and neighborhood change*. Chicago: Aldine.

Van Valey, T. L., Roof, W. C., & Wilcox, J. E. (1977). Trends in residential segregation, 1960-70. *American Journal of Sociology, 82*(4), 826-844.

Wienk, R., Reid, C. E., Simonson, J. C., & Eggers, F. J. (1979, April). *Measuring discrimination in American housing markets: The housing market practices survey.* Paper prepared for the Department of Housing and Urban Development, Office of Policy Development and Research.

4

Segregation in 1980: How Segregated Are America's Metropolitan Areas?

JOHN E. FARLEY

SCHOLARS CONCERNED WITH LEVELS OF housing segregation in America eagerly awaited the results of the 1980 Census, because this census offers the first real opportunity to assess the effects of the 1968 Open Housing Law. The first census after the passage of that law, taken in 1970, occurred too soon for any effects of the law to appear. By the 1980 census, however, the law had been in effect for about twelve years, more than enough time for any changes in segregation patterns resulting from the law to become evident. In the past few years, a number of analyses of housing segregation based on the 1980 Census have appeared in the sociological literature, and there is now enough evidence to draw some relatively firm conclusions about continuity and change in segregation patterns between 1970 and 1980. This chapter reviews that literature and draws conclusions about residential segregation patterns.

THE MEANING OF SEGREGATION

Segregation is defined as a tendency for members of different social (in this case, racial and ethnic) groups to live separately from one another, and with other members of their own group. The greater the extent to which members of any group live in areas with other members of the same group, and away from members of other groups, the more segregated the group is said to be.

It is also important to stress that, at least hypothetically, there are a number of social forces that can produce segregation. Although segregation clearly is often a result of discrimination, it cannot be automatically equated with discrimination. Segregation may reflect a preference to live with other members of one's own group, a lack of knowledge about housing opportunities away from one's own ethnic neighborhoods, or inability to afford living in areas occupied by other groups. In reality, the relative importance of these factors varies widely depending on the racial or ethnic group. This chapter will not address the causes of segregation, since other chapters in this volume focus primarily on that question. However, it will focus primarily on segregation of blacks, and, where evidence is available, Hispanics. As the other chapters show, it is segregation between these groups that appears to be most related to discriminatory processes.

CONCEPTS AND MEASURES OF SEGREGATION

Sociologists have used several ways to measure the basic concept of segregation, each of which taps a slightly different aspect. The most commonly used segregation index is the *index of dissimilarity* or D (Taeuber & Taeuber, 1965). This index, which ranges from 0 to 100, measures the extent to which any two groups are separated from one another. It can be computed for any large geographic area, such as a city or metropolitan area, which is divided into subareas such as blocks or census tracts for which data on racial composition are available. The meaning of this index can be illustrated with the example of black-white segregation. If a city had the maximum possible value of 100, or complete segregation, every black would live in a block or tract that is all black, and every white would live in a block or tract that is all white. If the city's index were 0, on the other hand, it would have no segregation. In this case, every block or tract would have exactly the same racial mix as the city as a whole. If the city were 20% black, so would be every block or tract within it. Actual values, of course, lie somewhere between 0 and 100, with the level of segregation being higher the closer the index comes to 100. In general, indices computed using block data are higher than indices for the same area using tract data, because they can detect segregation within the census tracts. (The formula for computing this index can be found in Taeuber and Taeuber, 1965, p. 236.)

The index of dissimilarity (D) is designed to give a measure of the degree of segregation of two groups regardless of the relative size of the two groups. This makes it an excellent unitary measure of overall segregation between any two groups, and readily allows comparison of segregation levels over time and among areas with different mixes. For some purposes, however, this strength can also be a weakness. It has been criticized, for example, on the grounds that it does not always give the best picture of the actual residential experience of the groups involved (Lieberson & Carter, 1982). In fact, the experience of two racial groups in a city with a given level of D can be quite different. Consider the example of a community of 10,000 blacks and 40,000 whites, where all of the blacks live on blocks that are 50% black and 50% white. In this community, 10,000 whites would live on such blocks, but the other 30,000 (75%) would live on all-white blocks. Thus whites would experience less exposure to blacks than blacks would to whites (Farley, 1984a).

For this reason sociologists have increasingly begun to use a second type of segregation index known as the P*-type or exposure index. This type of index is computed separately for each of the two groups involved, for example, a white-exposure-to-blacks and a black-exposure-to-whites index. These indices, which also range from 0 to 100, indicate the percentage of whites in the neighborhood where the average black lives, and vice versa for whites. Were there no segregation at all, the black exposure to whites would equal the percentage white in the city or metropolitan area, and white exposure to blacks would equal the percentage black in the area. These indices, unlike D, are influenced by the relative size of the two groups, but their advocates argue that for certain purposes this is appropriate, since this influences the amount of opportunities for interracial contact experienced by each group. (Formulae for computing this type of index can be found in Lieberson & Carter, 1982; Schnare, 1977; Farley, 1984a.) A related measure that is sometimes used is the percentage of blacks and whites who live on racially homogeneous blocks, for example, percentage of blacks living on blocks more than 80% black.

Segregation can also be evaluated by analyzing the patterns of blacks and Hispanics living mainly in the central cities, while whites live primarily in the suburbs. In some metropolitan areas, there has in the past been almost total exclusion of blacks from the suburbs. Although it is important to note that the absence of this type of segregation does not imply the absence of the types discussed just above (suburban blacks can and often do live in all-black neighborhoods), this type of city-suburb segregation has taken on a special importance in the social science literature, because to the extent that minority groups are absent from the suburbs, they are also separated from newer and better-quality housing, higher-achieving public schools, and increasingly suburban job opportunities.

SEGREGATION INDICES: 1980 AND EARLIER

In this section we shall focus on those that have used D and/or P*-type indices; in the following section we shall discuss city-suburb segregation. The largest number of these studies use D (Darden, 1985; Farley, 1982, 1984b; Hwang & Murdock, 1982, 1983; Logan & Schneider, 1984; Taeuber, 1983; Taeuber, Monfort, Massey, & Taeuber, 1984), though a number also use P*- or similar exposure-type measures (Farley, 1984a; Taeuber, 1983; Taeuber et al., 1984; Winsberg, 1982). Although a complete list of segregation indices for all metropolitan areas and central cities comparable to those published for 1970 (Schnare, 1977; Van Valey, Roof, & Wilcox, 1977) has yet to appear, the literature cited above covers a sufficient number of cities and metropolitan areas to begin to make some generalizations. We shall first consider black-white segregation, then review the more limited but growing evidence concerning Anglo-Hispanic and black-Hispanic segregation. Each discussion will begin with a brief overview of trends prior to the 1970s.

THE TREND IN BLACK-WHITE SEGREGATION BEFORE THE 1970s

Taeuber and Taeuber (1965) conducted an extensive literature review and analysis of data for the time period 1940-1960. In general, this revealed an upward trend in segregation levels in most parts of the country in the 1940s, followed by a decline in the 1950s. However, the changes were relatively small, and levels of segregation in most cities were very high in 1960.

When changes in segregation are small, as they were in the 1960s, it is difficult to state precisely the direction of change, because different measures may suggest different trends. Thus Sorensen, Taeuber, and Holingsworth's study of 109 central cities using D computed with block data revealed a modest decline in the average amount of segregation, while Van Valey et al. (1977), using D computed with tract data for all SMSAs for which such data were available in both 1960 and 1970, found no significant change in average segregation levels. The latter study did find that the average SMSA for which data were available in 1970 had a lower level of D than the 1960 average, but the reason for this was that SMSAs newly tracted in 1970 had lower levels of D than SMSAs for which data were available in both years. Thus when they limited themselves to a constant sample of the same cities, they found no appreciable change, except in the West, where there was a slight decline. Clearly, the differences between these two studies reflect their use of different samples of cities, and different levels of data (tract versus block) to compute their indices. Schnare's (1977) findings based on exposure indices for metropolitan areas further confirm this: She found that black exposure to whites decreased slightly while white exposure to blacks increased slightly. Since an average increase in the percentage black of metropolitan areas (which did occur) would by itself produce changes of this type, this study certainly indicates no abatement of segregation during the 1960s, and its author suggests that the findings may actually indicate an increase in the intensity of segregation.

To summarize, then, different studies produced slightly different results, but it is clear that overall, there was relatively little change in racial segregation levels during the 1960s, leaving segregation very high in 1970.

NEW FINDINGS: THE TREND IN BLACK-WHITE SEGREGATION DURING THE 1970s

The studies available thus far for 1980 suggest that, with some variation from city to city, there was on average a modest decline in segregation levels. In most cases the decline was small, although there does seem to be a clear downward trend in the 1970s. Nonetheless, most of the cities that have been studied thus far remain distinctly segregated.

The broadest national data available thus far pertain to large cities and metropolitan areas. Taeuber (1983) computed both D and exposure indices using block data for 28 large central cities, and Taeuber et al. (1984) computed D and exposure indices using tract data for 38 large metropolitan areas. In both cases, overall segregation, as measured by D, declined on the average. The average decline in the central city sample (using block data) was 6 points, from 87 to 81, while the average decline in the metropolitan area sample (using tract data) was 9 points, from 80 to 71 (keep in mind that tract data produce lower levels of D than block data). Clearly, despite the decline, the 1980 levels of segregation in these studies reveal a continued very high level of segregation in most U.S. cities and metropolitan areas. In at least some large central cities, such as Chicago, Cleveland, and St. Louis, there was no decrease at all in D; in one-fourth of the cities, D failed to decline by more than 2 (Taeuber, 1983; for greater detail on one such city, St. Louis, see Farley, 1982).

The exposure indices computed in these studies, though not as readily comparable to earlier years, also indicate continuing high levels of segregation. The white majority, in particular, remains quite unexposed to blacks in most metropolitan areas, with areawide nonblack-to-black exposure indices running below 5 in about half the metropolitan areas (Taeuber et al., 1984). In nearly all areas with sizable black populations, this index is well below half the area's percentage black. Black exposure to nonblack is higher, as we would expect given the large majority of whites in most metropolitan areas. However, when one makes the relevant comparison between this indicator and the percentage nonblack (mostly white), the typical exposure level in areas with sizable black populations is again well below half the areawide nonblack percentage. In some of the more segregated areas, such as Chicago, Detroit, and Cleveland, it is as low as one-quarter of that percentage. Comparisons with Schnare's 1970 figures generally suggest modestly higher intergroup exposure levels in 1980, although the measures used by the two studies are not identical. Taeuber (1983) also computed exposure indices using block data for his sample of 28 cities with black populations of 100,000 or more. These indices are harder to compare, since (1) the racial mix in these cities varies more widely than that of metropolitan areas, and (2) comparable indices are not available for many of the cities for 1970. Again, however, these indices indicate strikingly low levels of interracial exposure, with nonblack-to-black exposure ranging from 5 to 34, and black-to-nonblack exposure ranging from 7 to 31. These measures averaged 11.9 and 17.5, respectively, far below the level of exposure that would occur in the absence of segregation.

Hwang and Murdock's (1982) analysis of segregation in Texas cities, which included a wider range of city sizes than were covered by the studies cited above, used D computed with both block and tract data, and were based on data for both persons and householders. Indices were computed for 327 cities in SMSAs or other tracted areas, but most of the data reported are for the state's 38 central cities. The findings of this study indicated a somewhat larger average decline in D of about 12 (black-white segregation using block householder data) to 14 (black-Anglo segregation using tract person data). The 1980 indices averaged lower than those in Taeuber's samples; in the low 60s when computed with tract data, and around 80 when computed with block data. This suggests that black-white (or black-Anglo) segregation may be lower in smaller cities and/or areas with sizable Hispanic populations. The latter observation is also supported by the Taeuber et al. (1984) findings concerning metropolitan areas.

Finally, several studies suggest a decline in the proportion of blacks and whites living in racially exclusive areas between 1970 and 1980. Winsberg (1982), for example, found that the percentage of blacks living on blocks more than 80% black in ten Florida cities fell from an average of about 86% in 1970 to 68% in 1980. Similarly, Farley (1984a) found that the percentage of suburban St. Louis whites living in census tracts with less than 1% blacks declined from over 70% to about 52% during the same period, and that the percentage of blacks living in tracts 95% or more black fell from 24% to 16%. Significantly, however, the percentage of whites living in tracts 1%-5% black rose, as did the percentage of blacks in tracts 80%-95% black, and there was no real change between 5% and 80% black. Thus a decline in racial exclusivity does not imply changes in segregation at other levels of racial mix. Though he did not make comparisons to 1970, Taeuber's analysis of 28 central cities also included a measure of racial exclusivity. He found that in 20 of the 28 cities, the majority of nonblacks lived on blocks where at least one black was present. Similarly, in all but four of the cities, fewer than 30% of the blacks lived on all-black blocks. This is, of course, the most stringent possible measure of racial exclusivity, and it is clear that most blacks and whites live in areas *predominantly* composed of people of their own racial group.

FACTORS ASSOCIATED WITH SEGREGATION AND CHANGE IN SEGREGATION

These and other studies also indicate some of the factors that are associated with variation in black-white segregation levels. In general, it appears that the highest levels of segregation are found in the Midwest, with the lowest levels in the West. The Northeast and South occupy an intermediate position. In the Taeuber et al. (1984) study, for example, seven of the ten most segregated metropolitan areas were in the Midwest, and eight of the ten least segregated areas were in the West. Most of the areas with significantly below-average indices of dissimilarity (i.e., 65 or lower) also have relatively small black populations. The ten least segregated metropolitan areas in the Taeuber et al. (1984) study, for example, had an average percentage black of only 4.2%. These areas with low 1980 indices also typically experienced considerably greater than average downward change in segregation levels during the 1970s. Thus the areas with the lowest current levels of segregation also appear to be the ones that experienced the greatest downward change between 1970 and 1980. Finally, the majority of these areas have sizable Mexican-American populations, as do the areas included in the Hwang and Murdock (1982) analysis of Texas cities.

A distinctive feature of the Hwang and Murdock study was a regression analysis of the segregation levels. For black-white and black-Anglo segregation, the analysis indicated significant upward effects on segregation of percentage black, and (contrary to the foregoing discussion) percentage Hispanic. However, an interaction term of percentage Hispanic x population had some downward effect. This tentatively suggests that, in the cities in the national sample (Taeuber et al., 1984), it may be the low percentage black, not the high Mexican-American population, that produces the low segregation indices.

METROPOLITAN AND SUBURBAN VERSUS CENTRAL CITY SEGREGATION

Several studies also suggest that metropolitanwide and suburban segregation indices may be showing somewhat more evidence of decline than central city indices. In the two national studies, metropolitanwide segregation indices fell more than central city indices. This may be partly the result of using two studies with different samples of cities, but there is also some basis to believe that the difference is real. Farley's (1982, 1984a, 1984b) case study of St. Louis, for example, yielded similar findings: D (computed using block data) fell for the two of the three suburban counties with sizable black populations, but remained steady in the central city. The same two suburban counties also had indices below that of the city. On the other hand, there was very little change in exposure indices within either the city or suburbs, although black exposure to whites in the metropolitan area did increase slightly due to black suburbanization.

This study suggests two reasons that may account for a greater decline in metropolitanwide segregation than in central city segregation. First, segregation within suburban areas—at least the type measured by D—may be declining faster than segregation within central cities. Logan and Schneider's (1984) national analysis of segregation in suburbia does indicate a clear downward trend in D in most suburban areas during the 1970s, although because they used suburbs rather than tracts or blocks to compute D, it is not possible to compare the rate of decline with the central city studies. The second—and more likely—possibility is that metropolitanwide segregation indices are falling because an increasing proportion of blacks are moving to suburbia, which, though still segregated, is less segregated than the central city. We have already seen that this was the case for St. Louis, and other studies indicate the same for other areas. Another study by Hwang and Murdock (1983), for example, found that the average level of D for both blacks versus whites and blacks versus Anglos in Texas metropolitan suburbs was 73.5, compared to the central city averages of 80.3 for blacks versus Anglos and 79.3 for blacks versus whites. A case study of Kansas City conducted by Farley (n.d.) for use in school desegregation litigation indicated the same, with a difference of fully 27 points in D among the city and its Missouri suburbs (though blacks are badly underrepresented in those suburbs, as is the case in many of the Texas SMSAs studied by Hwang and Murdock). What does seem clear from these studies is that blacks are modestly less segregated within suburbia than within central cities, though (1) they remain underrepresented there and (2) some of the apparent low segregation indices may reflect "temporary integration" in racially changing suburban areas. These two issues will be explored later in the part of the chapter dealing specifically with black suburbanization.

HISPANIC SEGREGATION

Studies based on both 1980 data and earlier censuses are more limited than studies of black-white segregation. Until 1980, the census question relating to Hispanic origin was a sample question rather than a complete count question, which precludes use of the item at the block level. The studies that have been done generally indicate that Hispanics are less segregated from white Anglos than are blacks. However, studies from 1970 and earlier—nearly all using D as the measure of segregation—have indicated considerable variation among cities, regions, and Hispanic groups. In general, the highest levels of segregation between Anglos and Hispanics for years earlier than 1980 have been found for Puerto Ricans. Studies based on 1970 census data, for example, found Puerto Ricans to be nearly as segregated from white Anglos as blacks in Boston, Cleveland, and Seattle (Guest & Weed, 1976; Kantrowitz, 1979, pp. 49-52), and more segregated than blacks in Philadelphia (Hershberg, Burstein, Ericksen, Greenberg, & Yancey, 1979). The high levels of Puerto Rican segregation may in part reflect black-white segregation, since some Puerto Ricans are black or racially mixed; also their level of segregation may reflect the generally high levels of segregation found in large northeastern cities where the bulk of the Puerto Rican population resides. Analyses of segregation between Anglos and Mexican-Americans (Chicanos) have indicated lower average levels of segregation, but also more variation from city to city. In 1960, for example, the average level of D for Chicanos and Anglos in 35 southwestern cities was 54.5, compared to 80.1 for blacks and whites. Chicano-Anglo segregation was particularly high in Texas, where eight of the ten most segregated cities in the sample were located and had indices ranging from 63 to 76 (Grebler, Moore, & Guzman, 1970; Moore & Mitterbach, 1966). The average for all Texas cities was 59.3. Research by Lopez (1981) indicates a decline in segregation in Texas (as well as other parts of the Southwest) between 1960 and 1970, however. By 1970, the average level of D for Anglos and Hispanics (overwhelmingly Mexican-Americans) in 27 Texas cities was down to 49, still well above the figure of 43 for the entire Southwest and 35 for California (Lopez, 1981). Guest and Weed's (1976) findings for Cleveland and Boston suggest a high level of Anglo-Chicano segregation in 1970 in the Midwest and Northeast. In both cities, the index of dissimilarity between Chicanos and the balance of the population fell in the 1960s.

Particularly in 1970, and to a lesser extent in 1960, there is evidence that Chicano-black segregation levels were above Chicano-Anglo segregation levels in most southwestern cities. In 1970, the average Chicano-black index of dissimilarity in the Southwest was 55, or 12 points above the average Chicano-Anglo index (Lopez, 1981).

The areas covered by studies of Hispanic segregation based on 1980 data remain too limited to draw firm nationwide conclusions, but the studies available do point to relatively low levels of Chicano-Anglo segregation in the areas where they have been done compared both to past studies of Hispanic segregation and to levels of black-white segregation. Hwang and Murdock (1982) found a slight further decline in Anglo-Hispanic segregation in the 27 Texas cities studied by Lopez to an average of 43 (D based on tract data), and an average index of 38 for a wider sample of 38 cities. However, this decline was not uniform, and a few cities experienced increases in Anglo-Hispanic segregation. Even so, these levels of segregation are low enough that the averages are similar to levels reported by Guest and Weed (1976) and Kantrowitz (1979) for first- and second-generation European immigrant groups, and thus may in part reflect voluntary self-segregation (see also Lopez, 1981, p. 55). The index of dissimilarity for Anglos and Hispanics computed with block data in the larger sample averaged around 60—still a substantial level of segregation, but 20 points below the level of black-white segregation reported in the same study. Farley's (n.d.) case study of Kansas City shows that within the central city, where most of the area's Hispanic (predominantly Chicano) population lives, D based on tract data was 43, a level fairly typical of those found by Hwang and Murdock in Texas. In spite of the consistency of these studies, firm conclusions about 1980 levels and trends in Anglo-Hispanic segregation must await studies on Anglo-Chicano segregation in other southwestern and midwestern urban areas, as well as studies of segregation among Puerto Ricans and other Hispanics in the Northeast, and Cuban-Americans in Florida.

A final significant feature that emerges from the Hwang and Murdock (1983) and Farley (n.d.) findings is that Mexican Americans apparently continue to be more segregated from blacks than they are from whites. In the Texas sample, the average level of Hispanic-black segregation was 50 based on tract data and 76 based on block data—each well above the comparable figure for Hispanic-Anglo segregation, though there is considerable city-to-city variation. In Kansas City, the

Hispanic-black index was 75 in both the central city and the entire SMSA, almost as high as the black-white index and far above the Anglo-Hispanic index. The Texas averages did, as with other combinations, decline significantly between 1970 and 1980. Still, the fact that Hispanic-black segregation in 1980 exceeds Hispanic-Anglo segregation in most cities where the two kinds of segregation have been measured is another testimony to the special pattern of racial isolation experienced by black Americans.

CITY-SUBURB SEGREGATION

Perhaps the greatest change in segregation patterns in a number of metropolitan areas has been increased movement of blacks and Hispanics to the suburbs during the 1970s. Both groups have historically been underrepresented in the suburbs, and in the case of blacks, there was virtually no suburbanization during the 1960s despite rapid suburbanization of the larger population. The percentage of blacks living in the suburbs change only from 15.0% to 16.1% between 1960 and 1970, but by 1980 it had risen to 20.5% based on constant 1970 city boundaries (Long and DeAre, 1981). In addition, the percentage of blacks living in the central city declined for the first time after a steady rise lasting many decades. Thus the 1980s clearly represent a change in city-suburb segregation for blacks. As one might expect, the characteristics of suburbanizing blacks have become more similar to those of suburbanizing whites as black suburbanization has accelerated (Frey, 1984). With 37% of their population living in the suburbs (U.S. Bureau of the Census, 1982, p. 15), Hispanics are even more suburbanized than blacks, and this 1980 figure represents a continuation of a suburbanization trend that was already under way, though at a slower pace than white suburbanization. These figures suggest that the suburbs are more open to blacks and Hispanics, yet both groups remain underrepresented in suburbia, since 48% of whites were living in the suburbs by 1980 (U. S. Bureau of the Census, 1982, p. 15). In addition, the proportion of movers who go to the suburbs remains higher for whites than for blacks in most urban areas (Frey, 1984).

The extent of black suburbanization, in particular, varies widely among different metropolitan areas. In a few metropolitan areas, such as Washington, Los Angeles, and St. Louis, close to half the black population lives outside the central city. In Atlanta, nearly 40% does.

Even in these areas, blacks are quite underrepresented in the suburbs, but a substantial proportion of the area's blacks do live in the suburbs. In each of these areas, the percentage black in the suburbs, the percentage of the area's blacks living in the suburbs, and the absolute number of blacks in the suburbs all rose considerably.

There are, however, two sets of cities that present a sharp contrast to this group. The first is a group of mainly midwestern cities where blacks have been and continue to be extremely underrepresented in the suburbs. Examples of this group include Detroit (central city 63.1% black, suburbs 4.5% black), Chicago (central city 39.8% black, suburbs 5.6% black), Kansas City (27.4% black, Missouri suburbs 1.3% black), and Gary (central city 70.8% black, balance of SMSA 3.8% black). A few places—Gary, Detroit, and the Illinois part of the St. Louis area surrounding East St. Louis are the most notable examples—are well on the way to becoming economically depressed, overwhelmingly black cities, surrounded by relatively affluent, overwhelmingly white suburbs.

The second set of exceptions to the general rule of black suburbanization involves a number of southern metropolitan areas, including Memphis, New Orleans, Houston, and Dallas. These areas had a nominally sizable black suburban population in the past. The percentage black in the parts of these SMSAs outside the central city, for example, ranged from 8.3% in Dallas to 40.2% in Memphis in 1960 (Long & DeAre, 1981). The reality, however—particularly in Memphis and New Orleans—is that much of this population was poor and rural, but within counties included in SMSAs (see Logan & Schneider, 1984; Long & DeAre, 1981). Both the movement of affluent, suburbanizing whites into these areas and the mechanization of agriculture led to the displacement of this black population, with the result that in a number of southern metropolitan areas, the black population outside the central city either fell or grew more slowly than the white population. Thus while most metropolitan areas did experience some black suburbanization during the 1970s, there are important exceptions in both the North and South.

BLACK SUBURBANIZATION: HOW MUCH INTEGRATION DOES IT BRING

As noted above, several studies based on 1980 census data do suggest that there are modestly lower levels of segregation in the suburbs than in the central city. Even so, it is not clear whether there is any greater decline in suburban segregation than in central city segregation. In addition, suburban levels of segregation, although lower than that found in the central city, are still relatively high. Hwang and Murdock (1983), as noted, found an average D (based on tract data) for blacks and whites of 73.5, and Farley (1982) found D (based on block data) to range from around 80 to over 90 in three suburban counties around St. Louis. Logan and Schneider (1984) computed D using suburbs rather than tracts or blocks as the unit of analysis for 44 large metropolitan areas, an approach that detects segregation between, but not within, suburbs. Even so, 10 of the 44 metropolitan areas—mostly in the Northeast and Midwest—had indices above 60. With the exception of the Northeast, this index did decline in most areas. They also report the correlation ratio, a measure based on the degree of disparity in the percentage black in various suburbs. They note that this indicator shows less evidence of decline, especially in the Midwest and Northeast, suggesting that black population growth was greatest in the suburban areas that already had sizable numbers of blacks. If correct, this suggestion holds considerable implications for the meaning of black suburbanization: If blacks moving to the suburbs move mainly to areas where there are already established black neighborhoods, it is likely that black suburbanization will produce new "suburban ghettos," and a pattern of segregation much like that in the central city in earlier decades.

There is, in fact, a good deal of evidence that this is exactly what is happening in many parts of the country. In addition to computing segregation indices, Logan and Schneider used scatterplots and equation-fitting to judge the extent to which suburbanizing blacks during the 1970s moved mainly to suburbs that already had significant numbers of blacks present. They found that this was generally true in the Northeast and Midwest. Winsberg (1982) analyzed segregation in Florida, where Taeuber et al. (1984) found that a number of cities have segregation indices more typical of the Midwest than of the South. This study produced an additional important finding: Where there was rapid black population growth in predominantly white census tracts, it usually took place in ones contiguous with tracts that were 50% or more black. Farley's (1982) analysis of the St. Louis area, which considered segregation both within and between suburbs, produced similar findings. Most of the black population growth in the suburbs occurred in a sector that represented an extension of the area's traditional sector of black population, though parts of that sector extended quite far from the central city, and had been nearly all white in 1970. Although segregation was relatively low (D = 50 or lower, and sometimes around 30) in a number of suburbs within this sector, nearly all such suburbs experienced rapid growth in the black percentage. Suburbs with a D of 50 or lower experienced an average absolute change of 27 points in the percentage black. Thus suburbs that appear to be integrated in a one-time "snapshot" are more likely in transition from predominantly white to predominantly black. Outside this sector, there was little change in the black suburban population, and most suburbs that had any black population had very high levels of internal segregation. Considering together the Logan and Stearns, Winsberg, and Farley studies, it appears that black suburbanization is largely reproducing, in slightly milder form, the patterns of segregation previously found in the central city in most large, racially heterogeneous metropolitan areas in the Midwest, Northeast, and Florida.

In the South and West, on the other hand, suburban segregation declined in many areas. However, Logan and Schneider's (1984, pp. 882-885, 887) analysis indicates that much of the decline of segregation between suburbs in the Sunbelt is the result of a combination of white displacement of blacks in parts of suburbia (perhaps along the lines observed by Long & DeAre, 1981) and entry of suburbanizing blacks into other parts of suburbia. If such a trend continues, the recent trend toward reduced suburban segregation in the Sunbelt could be reversed, as some currently racially mixed suburbs move toward becoming predominantly white and others move toward becoming predominantly black. Thus although black suburbanization is probably producing modest declines in metropolitanwide segregation levels at present in most parts of the country, there is no guarantee that it will produce major declines from today's still high segregation levels, even in the West and South where declines have been most evident to date.

POSSIBLE FUTURE TRENDS IN SEGREGATION

Speculating about the future is always hazardous. Rarely do trends simply continue on a straight-line basis, and there are nearly always unforeseen events and changes that can cause even the best forecast to be wrong. Nonetheless, the trends in segregation, and the findings of studies concerning its causes (some of which appear elsewhere in this volume), do suggest some possible future trends in segregation, which we shall briefly consider here. With respect to black-white segregation, it is clear that there was a modest reduction in segregation levels between 1970 and 1980, though in most areas the level of segregation remains quite high. The decline in racially exclusive areas was particularly pronounced, suggesting that overt efforts to exclude blacks may be less widespread than in the past. This is consistent with survey data indicating that most whites, in the abstract, have no objection to living near a black of similar income and education. This means that blacks who are truly determined to move into all-white areas can probably do so. It probably also means a continuation of a slow reduction of segregation levels in most areas in the future.

However, for several reasons, it is not likely in the present social and political climate that there will be any rapid or dramatic decline in segregation levels. First, the predominant preference among whites continues to be to live in all-white or nearly all-white neighborhoods—in part because they fear racial change in the future of any neighborhood with more than a token black presence. Blacks, on the other hand, tend toward a preference for racially mixed neighborhoods. With this combination of preferences, white fears about neighborhood change in mixed neighborhoods become a self-fulfilling prophecy: Once a neighborhood becomes mixed, whites stop moving into it. Thus most of the people moving in will be black, which will, over the long run, tend to make the neighborhood predominantly black. Unless both blacks and whites continue to move into areas that are racially mixed, the areas cannot remain racially mixed. To break this cycle, two things would have to happen. First, more whites than now would have to develop a preference for racially mixed neighborhoods. Second, whites would have to recognize that their fears of racial change—and their resultant behavior—are part of the reason that racially mixed neighborhoods often do not remain so. At present, there is little evidence of either of these things happening on a large scale.

Besides the matter of white preferences and beliefs, there is the matter of racial steering in the real estate industry, which is discussed at greater length in other chapters in this volume. It is true enough that not all whites have the "typical" attitudes and beliefs described in the preceding paragraph, even though most do. Clearly, some whites are willing to move into racially mixed neighborhoods, or even desire to do so. However, the steering of such whites toward all-white neighborhoods by real estate agents makes it difficult for them to move to mixed neighborhoods. To most, the willingness or desire to move to a mixed neighborhood is not strong enough to lead them to resist a realtor's suggestions, and many remain unaware of housing opportunities in mixed areas. Thus as long as racial steering remains common, the prospects for a substantially accelerated decline in black-white segregation are not good.

It is true that black suburbanization increased substantially between 1970 and 1980, and there is reason to believe that this will continue to be the case, particularly if the middle-class segment of the black population grows further. However, as we saw, this by no means implies a great decline in segregation, since (1) segregation within the suburbs is only slightly less intense than within central cities and (2) the forces perpetuating segregation are as much at work in the suburbs as in the city.

The influence of racial steering raises another important question: How does political environment relate to the level of segregation? Racial discrimination in the sale and rental of most housing has been illegal since 1968, yet all evidence is that racial steering persists, and it is certainly clear that black-white segregation has at best declined only modestly. The answer appears to be that, while Open Housing legislation has greatly inhibited open discrimination ("I won't sell to you because you're black") by giving strong legal recourse to the victim, it has done little to inhibit subtler forms of discrimination. Indeed, it could do so only if governments were willing to commit substantial fiscal and human resources to a vigorous enforcement program. Most of the time, home buyers and renters never know that they have been subjected to racial steering, because they have no way to know what they have not been shown or told about. Only through audits where socioeconomically similar blacks and whites visit the same realty offices a short time apart can racial steering be verified. To conduct such audits regularly and nationwide on a scale sufficient to enforce the law would involve a commitment beyond what any administration thus far has been willing to make. The general low priority given by the Reagan administration to civil rights law enforcement, and the pressures against any costly undertaking arising from the massive federal deficit in the 1980s both suggest that no such commitment is likely in the near future. In the absence of such a commitment, racial steering is likely to continue (in part because realtors perceive that they are following the wishes of the majority of whites, which they probably are).

Another factor that limits the effectiveness of the 1968 Open Housing legislation is that it can do little to inhibit action that may be racially motivated but is claimed to be economically motivated. Racial factors have clearly played a role in local government decisions to block the construction of subsidized housing in predominantly white areas. However, this is clearly hard to prove, and economic considerations (which are perfectly legal) also play a role. Such actions by local governments are probably not as important as racial steering in perpetuating segregation, but they do clearly have that effect.

If a large decline has yet to come with respect to black-white segregation, it may be that it is largely past in the case of Hispanic-Anglo segregation. As noted, Hispanic-Anglo segregation indices in many cities are not only much lower than black-white indices, but are now comparable to those found for some "white ethnic" groups, and thus may in part reflect voluntary self-segregation. Unlike black-white indices, they are also often in a range that can in a substantial part be explained by socioeconomic differences from the white Anglo population. Higher indices do persist in some cities, but the preceding generalizations probably hold for the majority of cities for which Hispanic-Anglo segregation indices have been reported. These observations suggest that in many cities declining Anglo-Hispanic segregration may largely have run its course, since current levels are often explainable on the basis of group preferences and socioeconomics. Consistent with this interpretation is Hwang and Murdock's (1982) finding that, at least in Texas, the rate of decline in Hispanic-Anglo segregation was lower between 1970 and 1980 than between 1960 and 1970. For the reasons described above, it is likely that any further declines in Hispanic-Anglo segregation will be small—but it is important to keep in mind that Hispanic-Anglo segregation is already typically much lower (at least in the areas for which we have information) than black-white segregation, and that it has already declined a good deal more than black-white segregation.

For these reasons, any great or sudden decline in black-white segregation over the next few years appears unlikely, particularly in metropolitan areas with large black populations. It is, however, true that the present slow decline probably will continue. But Taeuber et al. (1984, p. 9) show that a continuation of the 1970-1980 trend for another 50 years would still leave the average central city with a D (based on block data) of above 50. The pace of decline may accelerate slightly, but it is not likely to accelerate dramatically. Thus our urban areas will likely remain quite segregated well into the future unless there is a major change in the social forces currently at work.

REFERENCES

Darden, J. T. (1985). *The significance of race and class in residential segregation.* Paper presented at annual meeting of the Urban Affairs Association, Norfolk, VA.

Farley, J. E. (1982). Metropolitan housing segregation in 1980: The St. Louis case. *Urban Affairs Quarterly, 18*, 347-359.

Farley, J. E. (1984a). P*-segregation indices: What can they tell us about housing segregation in 1980? *Urban Studies, 21*, 331-336.

Farley, J. E. (1984b). Housing segregation in the school-age population and the link between housing and school segregation: A St. Louis case study. *Journal of Urban Affairs, 6*(4), 65-80.

Farley, J. E. (n.d.). *Indices of residential segregation prepared for use in Kansas City school desegregation case.* Unpublished mimeograph, tables only.

Frey, W. H. (1984). Lifecourse migration of metropolitan whites and blacks and the structure of demographic change in large central cities. *American Sociological Review, 49,* 803-827.

Grebler, L., Moore, J. W., & Guzman, R. C. (1970). *The Mexican-American people.* New York: Free Press.

Guest, A. M., & Weed, J. A. (1976). Ethnic residential segregation: Patterns of change. *American Journal of Sociology, 81*, 1088-1111.

Hershberg, T., Burstein, A., Ericksen, E., Greenberg, S., & Yancey, W. (1979). A tale of three cities: Blacks and immigrants in Philadelphia. 1850-1880, and 1970. *Annals of the American Academy of Political and Social Science, 441*, 55-81.

Hwang, S. S., & Murdock, S. H. (1982). Residential segregation in Texas in 1980. *Social Science Quarterly, 63*, 737-748.

Hwang, S. S., & Murdock, S. H. (1983). Segregation in metropolitan and nonmetropolitan Texas in 1980. *Rural Sociology, 48*, 607-623.

Kantrowitz, N. (1979). Racial and ethnic segregation: Boston, 1830-1970. *Annals of the American Academy of Political and Social Science, 441*, 41-54.

Lieberson, S., & Carter, D. K. (1982). Temporal changes and urban differences in residential segregation: A reconsideration. *American Journal of Sociology, 88*, 296-328.

Logan, J. R., & Schneider, M. (1984). Racial segregation and racial change in American suburbs, 1970-1980. *American Journal of Sociology, 89*, 875-888.

Long, L., & DeAre, D. (1981). The suburbanization of blacks. *American Demographics, 3*(8), 16-21, 44.

Lopez, M. M. (1981). Patterns of interethnic residential segregation in the urban southwest, 1960 and 1970. *Social Science Quarterly, 62*, 50-63.

Moore, J. W., & Mitterbach, F. G. (1966). *Residential segregation in the urban southwest: A comparative study* (Mexican American Study Project, Advance Report 4). Los Angeles: University of California Press.

Schnare, A. B. (1977). *Residential segregation by race in U.S. metropolitan areas: An analysis across cities and over time* (Contract No. 246-2). Washington, DC: Urban Institute.

Sorensen, A., Taeuber, K. E., & Holingsworth, L., Jr. (1975). Indexes of racial residential segregation for 109 cities in the United States, 1940-1970. *Sociological Focus, 8*, 125-142.

Taeuber, K. E. (1983). *Racial residential segregation, 28 cities, 1970-1980* (CDE working paper 83-12). Madison, WI: University of Wisconsin-Madison, Center for Demography and Ecology.

Taeuber, K. E., Monfort, F. W., Massey, P. A., & Taeuber, A. F. (1984). *The trend in metropolitan racial residential segregation.* Paper presented at annual meeting of the Population Association of America, Minneapolis.

Taeuber, K. E., & Taueber, A. F. (1965). *Negroes in cities.* Chicago: Aldine.

U. S. Bureau of the Census. (1982). *Statistical abstract of the United States* (103rd ed.). Washington, DC: Government Printing Office.

Van Valey, T. L., Roof, W. C., & Wilcox, J. E. (1977). Trends in residential segregation: 1960-1970. *American Journal of Sociology, 82*, 826-844.

Winsberg, M. D. (1982). Changing distribution of the black population: Florida cities, 1970-1980. *Urban Affairs Quarterly, 18*, 361-370.

5

The Suburbanization Process and Residential Segregation

THOMAS A. CLARK

MANY OBSERVERS have found evidence of diminished discrimination (Goodman & Streitwieser, 1983), minority advance, and improved life prospects (Clark, 1979; Frey, 1978) in the accelerating pace of minority suburbanization during the late 1960s and 1970s. Between 1970 and 1980, the black population in suburbs increased by almost 50% (Guest, 1978; Long & DeAre, 1981; Schnore, 1976). The gain was driven mainly by migration, not natural increase, and the prime origin of black migrants was the central city (Spain & Long, 1981). About seven in every ten black migrants to the suburbs came from these cities in the 1970s. In fact, over one in ten blacks living in central cities in 1970 moved to the suburbs in the ensuing decade (Clark, 1979, exhibits 1, 8). The white rate of city-to-suburb movement, however, was double the black movement.

Still at issue, however, is whether suburbanization actually emerges from or enables minority welfare gain. Does the momentum in the movement of blacks to suburbs persist in the 1980s? Have other minorities won a suburban foothold paralleling or arising from the black experience? These concerns are the subject of this chapter. The case is built that the black trend established in the 1970s persists in the 1980s. Blacks continue suburbanward at a high rate. This trend and the internal pattern of black suburban destinations, however, vary widely among subnational and metropolitan regions. Where the flow is large and economically diverse and only a few suburban jurisdictions receive most of the migrants, the curve of welfare gains may soon arc downward as city-suburban welfare differences diminish. In these places, suburbanization may slow unless the income mix of black migrants becomes more skewed to wealthier households and a broader array of destinations emerges. But the trend will probably accelerate still more where the black migrant flow is currently high, the pool of middle- and upper-income households in the central city is large, and the number of suburban jurisdictions receiving blacks is substantial. Hispanic suburbanization appears to parallel but not replicate the non-Hispanic black experience. In some places, the two processes occur in tandem. In others, they occur independently but are energized by the same social and spatial realities (see also Jackson, 1985; Schwartz, 1980).

Overall, it will also be argued, both short- and long-run gains of minority migrants remain problematic as is the impact of city departures on those left behind. Minority preferences for residential "self-selection," as well as both "housing" and "spatial" discrimination—revealed in landlord and seller bias, and exclusionary proclivities of suburban jurisdictions, respectively (Clark, 1982; James & Crow, 1984; Lake, 1984)—have probably lessened more than marginally in most areas. Increase in the number and prosperity of the set of minority destinations in the suburban ring is here taken to be evidence of diminished residential discrimination. Decline in residential discrimination and increase in minority welfare, however, are not necessarily or immediately symmetrical. New nonresidential forms of denial of minority access may emerge in suburbs. These would dampen migrant gains. And the loss of more prosperous minority households can reduce the viability of central city economies and jeopardize the welfare of those who remain. Minority suburbanization has not always arisen from a lessening of suburban residential discrimination. It may, paradoxically, be promoted by economic reconfigurations in central cities that yield racial and ethnic displacements. The supposed benefit accruing to suburbanward migrants may be heightened by welfare loss in cities, not increase in the absolute benefit realized by minority suburbanites themselves.

NATIONAL TRENDS IN MINORITY SUBURBANIZATION SINCE THE MID-1970s

Since the mid-1970s, the effective annual rate of both black and Hispanic suburbanization—city-to-suburb migrants as a percentage of total central city population in each group—has accelerated significantly. Central cities continue to be the major source of *net* migrants to the suburban rings of metropolitan areas (SMSAs) for both groups. And the net contribution of nonmetropolitan areas to Hispanic suburban increase has been especially large.

In recent years, the Hispanic rate of net migration has exceeded the black rate to the suburbs. During the 1975-1980 and again in 1980-1983 net black migration (migrants *to* less migrants *from* the suburbs) were equivalent to 16.4% of the total suburban black population at the start of each interval (Table 5.1). Net migration from the central city was particularly strong. It accounted for a 10.8% increase in the black suburban population during 1975-1980, and 14.2% in 1980-1983. The nonmetropolitan contribution of black net migrants actually fell from 5.6% to 2.2% of the black resident population in each period. Net Hispanic migration to the suburban rings of SMSAs increased the Hispanic presence in suburbs by one-fifth in each interval: 21.6% in 1975-1980 and 20.8% in 1980-1983. Over half this increase in each interval was due to net migration from central cities: 11.2% and 12.3%, respectively. The contribution of net Hispanic migration from nonmetropolitan areas was inordinately high in each interval: 10.4% and 8.5%, respectively.

The rapid influx of both Hispanics and blacks in the suburban rings of the nation's metropolitan areas vastly outpaces that of whites (Table 5.1). In these data, the persons of "Spanish origin"—Hispanics—may be of any race. Most, however, are self-identified as nonblack. "White" includes some of Spanish origin. During the period 1975-1980, the white population in suburbs increased 7.8% as a result of net migration, and during the period 1980-1983 by 8.2%. These white rates fall far short of those of blacks and Hispanics. Moreover, if most Hispanics are "white," then almost one in four new net migrants to the suburbs were nonmajority during both intervals. The minority component of the net stream of migrants to suburbs from central cities rose from 15% during 1975 to 1980, to 22% in 1980 to 1983. Surprisingly, the minority element of the net stream from nonmetropolitan areas was almost 90% in the period 1975-1980, but fell to 33% in the second interval. These latter figures may marginally overstate the "minority" component of black plus Hispanic rates since all Hispanics were designated "white." Metropolitan areas in which a significant share of the Hispanic populace originated in the Caribbean region—along the Gulf and East Coasts—would be expected to have more black Hispanics.

Minority-majority differences in suburbanization rates, at least through the 1970s, have been largely attributable to noneconomic

TABLE 5.1
Movers To and From the Suburban Ring by Race and Spanish Origin[1], 1975 to 1983[2]
(Numbers in Thousands)

Period	White		Black		Spanish Origin	
	Number	*Percentage of Total*[3]	Number	*Percentage of Total*[3]	Number	*Percentage of Total*[3]
1975 to 1980						
Population in 1975	69,239	100.0	4,196	100.0	3,342	100.0
Movers to Suburbs:						
From Central Cities	8,840	12.8	915	21.8	716	21.4
From Outside SMSAs[4]	4,395	6.3	341	8.1	418	12.5
Movers from Suburbs:						
To Central Cities	3,928	5.7	463	11.0	342	10.2
To Nonmetropolitan Areas	3,968	5.7	104	2.5	69	2.1
1980 to 1983[5]						
Population on 1980	78,946	100.0	5,160	100.0	4,413	100.0
Movers to Suburbs:						
From Central Cities	9,942	12.6	1,331	25.8	1,033	23.4
From Outside SMSAs[4]	4,519	5.7	276	5.3	485	11.0
Movers from Suburbs:						
To Central Cities	4,807	6.1	597	11.6	492	11.1
To Nonmetropolitan Areas	3,141	4.0	158	3.1	109	2.5

SOURCE: Author's calculations based on U.S. Bureau of the Census, *Current Population Reports* (1981, 1983, 1983, and 1984, respectively).
1. Persons of "Spanish Origin" may be any race.
2. Time intervals commence and end in March.
3. Total population is nonmovers plus movers within suburbs and out-movers from suburbs, alive at the start of each time interval.
4. Includes in-movers from abroad.
5. This period is the sum of the 1980-1981, 1981-1982, and 1982-1983 flows.

discrimination, and a marginal minority propensity toward self-selection. To these add a growing gap between suburban and nonsuburban housing costs in the face of the superior buying power of the majority, white population. Controlling for SMSA structure, city decline, and nonracial characteristics of the population at risk, the overall black-white city-to-suburb migration rate differential has been found to stem more from black "retention" in central cities than from accelerated white "flight" (Goodman & Streitwieser, 1983). Actual or anticipated discrimination in suburban housing markets is uniformly asserted to play a major role in elevating minority retention outside the suburbs (James, McCummings, & Tynan, 1984; Wienk, Reid, Simonson, & Eggers, 1979). But differences in buying power (Clark, 1981; Frey, 1978; Hermalin & Farley, 1973, 1979; Nelson, 1979), age, and household composition of cities and nonmetropolitan areas (Frey, 1984), and mobility rate—fraction of residents moving per unit of time (Goodman & Streitwieser, 1983)—also help account for minority-majority differences in rates of out-movement to the suburbs.

SUBURBAN SELECTIVITY OF MINORITY MIGRANTS

The pace, pattern, and meaning of minority suburbanization vary widely among subnational regions and metropolitan areas. Twenty metropolitan areas have been identified, five from each major census region, to explore this variation. These twenty are first examined to assess the minority composition of their suburban rings and document the evolving trend in the "suburban selectivity" of blacks destined for suburbs from both central cities and nonmetropolitan areas. These are then contrasted against the comparable white trend since the mid-1960s and the Hispanic trend in the period since the mid-1970s for which appropriate census data are available. These data reveal differences among groups in the propensity to move to, or remain within, the suburban realm. They do not address differences in suburbanization rate or mobility rate.

DATA

Of the twenty cities identified for detailed study, five are drawn from each census region. Each is a metropolitan area (SMSA or "standard metropolitan statistical area") whose consolidated equivalent (CMSA) ranked among the top twenty in size in 1980, excepting Denver ranked twenty-first, and New Orleans ranked twenty-seventh. Newark is part of the New York CMSA. Minneapolis-St. Paul, Seattle-Tacoma, and San Diego ranked in the top twenty but are not included. The set contains the four largest within each region, plus one other.

The selectivity quotients shown in Figure 5.1 are reported for three periods for blacks and whites: 1965-1970, 1975-1980, and 1980 through the latest date for which data were available. The Hispanic quotients are reported only for the latter two intervals. Comparable Hispanic migration data were not available in prior years. All quotients are calibrated for "recent movers" who were alive at the start of the respective interval and survived to the end of the period.

Sources are documented in Figure 5.1-D. *Whites* are defined as nonblacks in all three periods. The metric is "persons" in the first and second, and "householders" in the third. The first and last intervals address suburban rings equivalent to the noncentral city portion of SMSAs defined in 1970, whereas the second is based on SMSAs defined in 1980. The number of householders equals the number of households. Quotients based on moving persons and households are presumed to be comparable. The Annual Housing Survey, the source of the data for the third interval, counts only migrant households that had the same head before and after the move. These amount to about three-quarters of all households.

DEFINITIONS

Unlike the "suburbanization rate," which is the fraction of a subpopulation residing in the central city that moves to the suburbs during a unit of time, "suburban selectivity" quotients measure the fractions of movers in cities, suburbs, and nonmetropolitan areas, respectively, who relocate to new suburban residences per unit of time. These quotients assess the combined effect on locational choice of household preferences and resources, in light of available suburban housing opportunities.

The central city's suburban selectivity quotient is higher than the suburbanization rate since its denominator is "movers" and not total population of which movers are a part (see also Frey, 1985). The decision to seek a new residence is governed more by changing residential preferences that is driven largely by life-cycle change, and changing resources with which to buy or rent, than by the perceived availability of alternative housing opportunities. The mobility rate therefore intercedes between the rate of suburbanization and the suburban selectivity quotient of residents in cities who have elected to move.

Below are the key quantitative definitions needed to interpret Figure 5.1.

Define for each time period, for whites, blacks, and Hispanics, respectively,

I = movers within the suburban ring,

T = movers from central city to suburban ring within the same metropolitan region (SMSA),

F = movers from suburbs to the central city within the same metropolitan region under study,

M = movers from other metropolitan regions to the suburbs of the metropolitan region under study,

U = movers from nonmetropolitan areas to the suburbs of the metropolitan region under study,

C = movers within the central city,

V = movers from other metropolitan areas to the central city of the metropolitan region under study.

Note that w, b, and s are subscripts denoting white, black, and Spanish origin or Hispanic, respectively.

Then, we have

suburban selectivity of:

(1) suburban movers $= I/(I + F)$

(2) city movers $= T/(T + C)$

(3) movers to SMSA $= (M + U)/(M + U + V)$

city selectivity of suburban movers (not shown in Figure 5.1) is given by

$$F/(I + F) = 1 - [I/(I + F)]$$

and,

> *net in-migration to the suburban ring from the central city by group, as a fraction of total in-migration from the central city which follows:*

(1) for whites during all intervals,

$$(T_w - F_w)/[(T_b - F_b) + (T_w - F_w)]$$

(2) for blacks during all intervals,

$$(T_b - F_b)/[(T_b - F_b) + (T_w - F_w)]$$

(3) and for Hispanics during 1975-1980, and since 1980,

$$(T_s - F_s)/[(T_b - F_b) + (T_w - F_w)]$$

INTERPRETATION

The results of the analysis of migrant suburban selectivity are presented in Figure 5.1 (A-D), displayed by region. Among regions, the higher levels of minority suburbanization—group population as a fraction of the total in suburbs—appear in the South and West. The black levels, denoted by the symbol "+," are particularly high in Washington, D.C., Miami, Atlanta, and New Orleans in the South; and elsewhere in Newark (Northeast), St. Louis (North Central), and Los Angeles-Long Beach (West). Hispanic suburbanization levels, shown by an asterisk in Figure 5.1, are inordinately low in the northern regions, but markedly higher in Miami in the South, and in Los Angeles-Long Beach, San Francisco-Oakland, and Houston in the West.

Consider the dynamic underlying these conditions by examining the suburban selectivity quotients in each region, respectively. In the Northeast, an overall trend toward the convergence of the suburban selectivity quotients of both blacks and whites who already reside within the suburbs is underway. In both New York and Newark, however, the selectivity of black suburban movers is declining. In Boston this quotient has increased dramatically among black movers, indicating

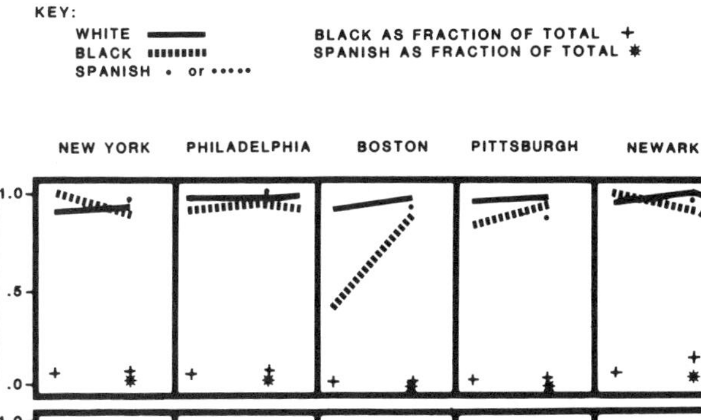

SOURCES: U.S. Bureau of the Census, 1973, Table 15; U.S. Bureau of the Census, 1984, *Census of Population, 1980,* Table 10; and U.S. Bureau of the Census, 1981, 2 and 3, *Annual Housing Survey*, Tables 3, 13 and 23.

Figure 5.1A Suburban Selectivity of Households Moving from Dwellings in Cities, Suburbs, and Nonmetropolitan Areas During 1965-1970, 1975-1980, and Since 1980: Selected Cities of the Northeast

much stronger retention of blacks already in residence. The white suburban selectivity quotient has remained high, as has the Hispanic, and steady.

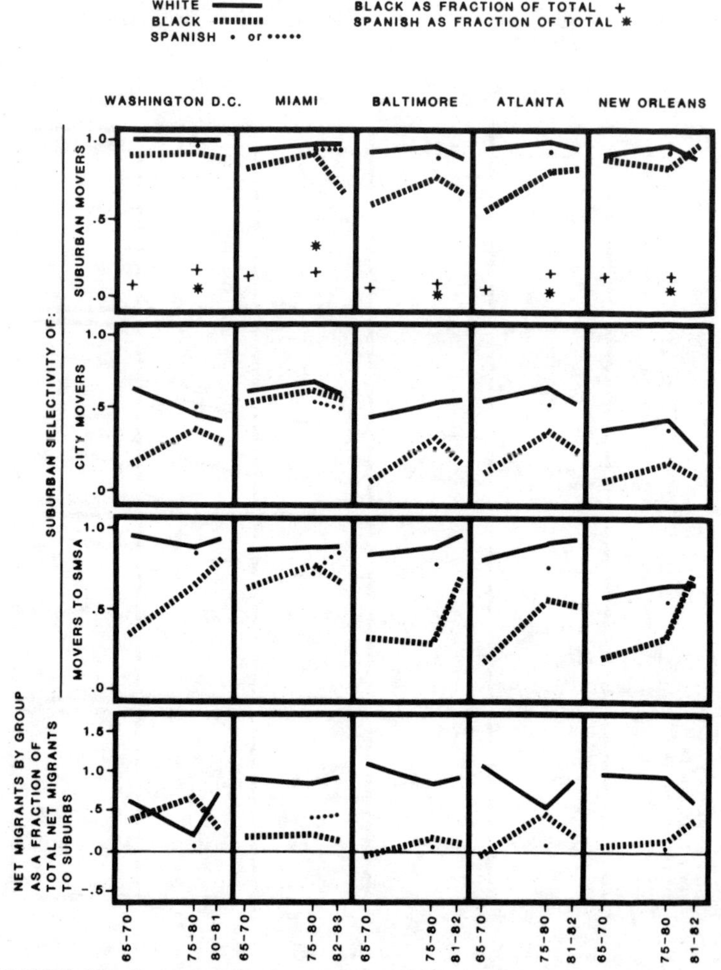

SOURCES: U.S. Bureau of the Census, 1973, Table 15; U.S. Bureau of the Census, 1984, *Census of Population, 1980,* Table 10; and U.S. Bureau of the Census, 1981, 2 and 3, *Annual Housing Survey*, Tables 3, 13 and 23.

Figure 5.1B Suburban Selectivity of Households Moving from Dwellings in Cities, Suburbs, and Nonmetropolitan Areas During 1965-1970, 1975-1980, and Since 1980: Selected Cities of the North Central Region

Not only is the central city now the prime source of suburban-bound migrants, but if current mobility rates persist or increase, the city's contribution to minority increase in the suburbs is likely to expand. Of

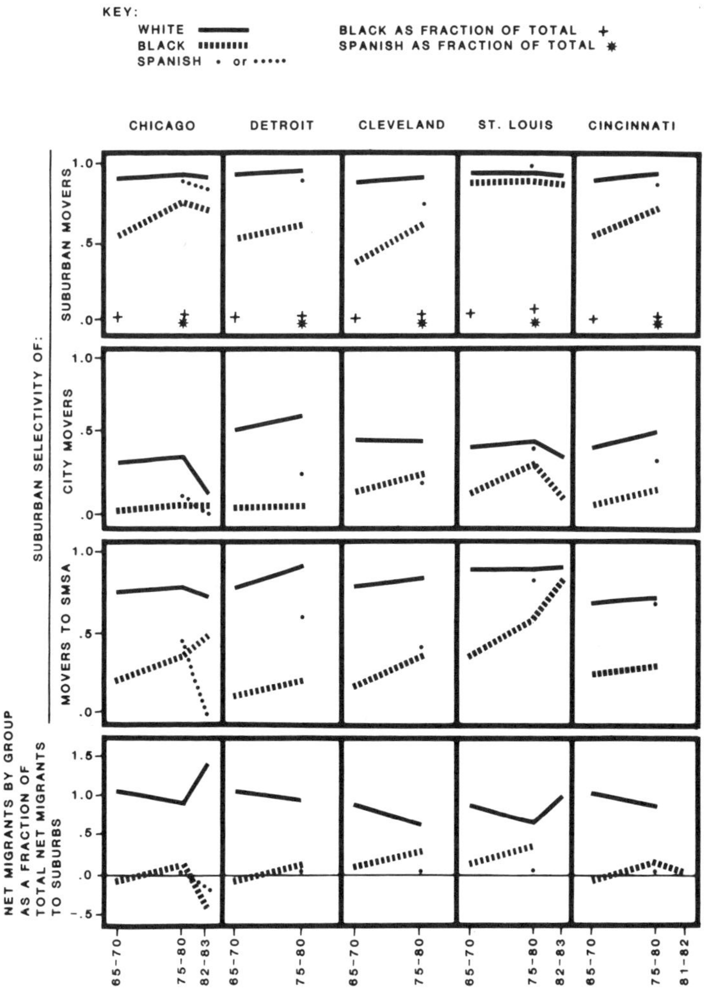

SOURCES: U.S. Bureau of the Census, 1973, Table 15; U.S. Bureau of the Census, 1984, *Census of Population, 1980,* Table 10; and U.S. Bureau of the Census, 1981, 2 and 3, *Annual Housing Survey*, Tables 3, 13 and 23.

Figure 5.1C Suburban Selectivity of Households Moving from Dwellings in Cities, Suburbs, and Nonmetropolitan Areas During 1965-1970, 1975-1980, and Since 1980: Selected Cities of the South

course, if the trend is carried to an extreme, the residential condition of minorities in central cities may improve, upon a reduced population base. Even constant rates of mobility and suburban selectivity, however,

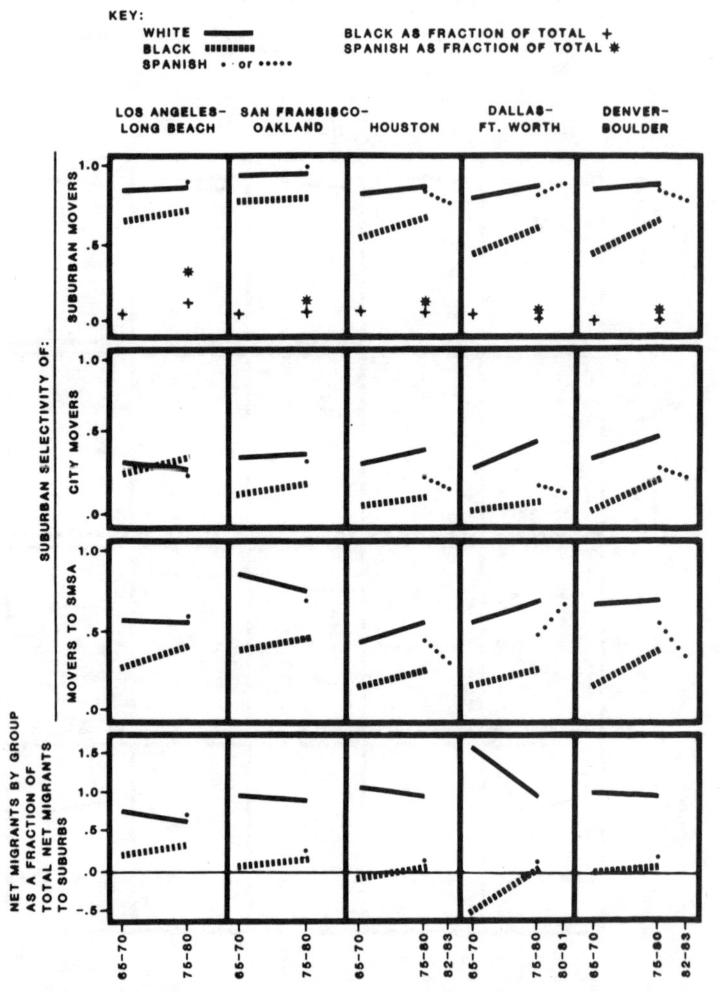

SOURCES: U.S. Bureau of the Census, 1973, Table 15; U.S. Bureau of the Census, 1984, *Census of Population, 1980,* Table 10; and U.S. Bureau of the Census, 1981, 2 and 3, *Annual Housing Survey*, Tables 3, 13 and 23.

Figure 5.1D Suburban Selectivity of Households Moving from Dwellings in Cities, Suburbs, and Nonmetropolitan Areas During 1965-1970, 1975-1980, and Since 1980: Selected Cities of the West

would fail to maintain the current levels of minority suburbanization if city minority population declines. In the sample central cities of the Northeast, white suburban selectivity is generally high and increasing

(Figure 5.1-A). Both New York and Newark, however, witness declining white selectivity quotients until after 1980. The suburban selectivity quotients of black movers in central cities remained low in New York and Philadelphia, but rose significantly in Pittsburgh and Newark. Hispanic quotients paralleled the black, at least during the period 1975-1980.

Minority migration to the suburbs from nonmetropolitan areas is a less frequently noted contributor to suburban increase, but it is likely to play an expanding role. Increase in the flow of minorities to the suburbs from nonmetropolitan areas, however, is contingent on the continued expansion of the suburban job base, and a slowing of the dispersal of employment including manufacturing into nonmetropolitan areas. These account for apparent positive differences between suburbs and adjacent nonmetropolitan areas in minority income potential. The same can be said of intermetropolitan differences. Among the sample metropolitan regions of the Northeast, white suburban selectivity rates of movers to metropolitan areas are high but undergoing modest declines. New York is the exception (Figure 5.1-A). There, central city housing is costly and in short supply relative to the suburbs, so white suburban selectivity is rising rapidly. The suburban selectivity quotients of blacks migrating from nonmetropolitan areas and other metropolitan areas are also rising in most metropolitan areas of the Northeast. Falling quotients in New York, however, may stem from the lower income mix of the black migrant stream from outside the New York region. Declines in suburban selectivity in Newark after 1980 may reflect a marginal strengthening of the city economy and the growing income exclusiveness of its suburbs. Hispanic quotients fell near the black in 1975 through 1980 except in Pittsburgh where they equaled the white.

The overall racial and ethnic configuration and mix of the resident population in suburbs is a prime conditioner of the entry experience of new in-migrants. Of equal significance, however, may be the racial and ethnic composition of the new immigrant flow itself. Where the number of suburban arrivals within a single race or ethnic group exceeds departures, the spatial extent of minority residence is bound to expand if only marginally. Of course, the degree of expansion and reconfiguration of suburban minority residential patterns as a result of the exchange of in- and out-migrants will be conditioned by the relative incomes, household identities, and job sites of arriving and departing households. Still, a one-to-one correspondence of the spatial patterns of residential departures and arrivals per unit time in suburbs should not be expected, even when the two flows have a common socioeconomic identity. In the Northeast, whites constitute a declining share of new migrants to suburban areas in all metropolitan areas examined, though a reversal is noted in Newark after 1980 (Figure 5.1-A). The black share actually rose from a negative position in Boston during the period 1965-1970 to 1975-1980, indicating the net black flow was initially out of the suburbs. As a result, the white share of new migration actually exceeded unity in this region in the period 1965-1970. In all northeastern metropolitan areas examined, the Hispanic share of net migration was nearly zero.

Among the sampled metropolitan areas of the North Central census region, the trend in suburban selectivity is strikingly similar to the Northeast's despite important structural differences in recent economic performance (Figure 5.1-B). Suburban selectivity of suburban movers in the five metropolitan areas examined exhibits high, steady white quotients. Rising black quotients signal improved retention of minority residents. St. Louis' black and white quotients are uniformly high though the black population is highly concentrated in spill-over communities just outside the central city. The Hispanic quotients parallel the white in all cases in 1975 through 1980. In contrast with the metropolitan areas of the Northeast, those of the North Central region exhibit considerably lower levels of retention of blacks living in suburbs.

The parallel between the cases of the Northeast and North central regions extends to the other measures of migrant selectivity. All measures, when tracked since 1965, demonstrate a growing convergence of black and white selectivities. In the North Central region, however, the black-white disparity at onset of this analysis was appreciably greater. Suburban selectivities of movers to metropolitan areas and of city movers exhibit racial convergence overall. Hispanic selectivities fall within the black-white range but are irregular. The summed effect of suburban selectivities in the North Central region is a marked evolution toward an actual convergence of the proportionate contributions of black and white net migrants to the suburbs in Cleveland and perhaps St. Louis. Elsewhere, the black contribution to net suburbanward migration is low, and actually declining in two cases: Chicago and Cincinnati. Chicago stands apart from the others. Whites there account for over 100% of the net migration to suburbs due to actual net black outflows since 1980. In Cincinnati the black contribution to net aggregate migration to suburbs is also in decline.

Some of the largest gains in minority suburbanization have occurred in the metropolitan areas of the South (Figure 5.1-C). There the suburban selectivity of suburban movers, both black and white, is high. Marginal declines, however, are noted since 1980 in black selectivity quotients in Washington, D.C., Miami, and Baltimore. In New Orleans a larger share of black than white resident suburban movers is apparently being retained within the suburban ring. Black city movers were far more inclined to move to the suburbs in all cases examined in the period 1975-1980 than in the period 1965-1970. But at least a marginal lessening in this inclination is observed in all cases since 1980. Strikingly high selectivity quotients of black and Hispanic moving to the metropolitan region from elsewhere are also noted. The contribution of blacks to net aggregate suburbanward migration are particularly high in Washington, D.C., Atlanta, and, since 1980, in New Orleans.

The West presents still another contrasting set of trends in suburban selectivity (Figure 5.1-D). Here the suburban selectivity of black movers residing in suburbs lags far behind the white. The Hispanic generally resembles the white since the mid-1970s, and this disinclination to depart helps account for the relatively large proportionate presence of Hispanic in the suburbs of Los Angeles-Long Beach, San Francisco-Oakland, and Houston. Ironically, the selectivity quotients of black city movers in most of the metropolitan areas of the West is low, and where measured, the Hispanic equivalent was in decline in the years since 1975. Los Angeles-Long Beach is the exception. Here the suburban selectivity of black city movers exceeds the white and, recently, the Hispanic. In addition, the suburban selectivity of black movers to the suburbs has been quite low relative to the counterparts of the Northeast and South. Still, the overall share of aggregate net migration to suburbs claimed by whites is declining in all areas of the West, but especially rapidly in Dallas-Fort Worth where the net flow of blacks to the suburbs was highly negative prior to 1970.

RACIAL DIFFERENCES IN RESIDENTIAL PATTERNS WITHIN SUBURBAN RINGS

Quotients of suburban selectivity identify the major components of the streams of migrants to and from the suburbs but do not address the internal configuration of residence within the suburban ring. The "dissimilarity index" is a measure that is conventionally used to express the degree of disparity between any two residential patterns (see Elgie,

1979; Sorensen, Taeuber, & Hollingsworth, 1975). Such indices have been assembled for each central city and suburban ring of the 20 metropolitan areas identified in earlier analysis, for both 1970 and 1980 (Table 5.2). The current application considers only black-white differences.

The dissimilarity index assumes values between 0 and 100, and indicates the minimum percentage of either whites or blacks that would have to move in order to achieve uniform distributions for each race across all spatial units examined. Typically, as for the central city indices reported in Table 5.2, the spatial units are census tracts (see Taeuber, 1983). Analysts also have employed tract data in suburban studies of dissimilarity (see James et al., 1984), but the "gate-keeping" function that determines ease of minority entry is more the province of municipalities than of neighborhoods in suburbs (Clark, 1981, 1982, 1984; Lake, 1981b, 1984; Logan, 1978; Logan & Stearns, 1981). Consequently, the suburban indices reported in Table 5.2 are for incorporated suburban jurisdictions (see Logan & Schneider, 1984).

The degree of dissimilarity across suburban jurisdictions of black and white residential patterns is one indicator of welfare gain realized through minority suburbanization. The greater the similarity of patterns, the less may be the level of discrimination encountered as minority households enter the suburbs. Welfare gain would be presumed greater when blacks, Hispanics, and others win access to the more preferred jurisdictions. These places have tended toward greater racial exclusivity arising partly from the high cost of living there (Grier & Grier, 1983; Lake, 1981a; Rose, 1976). Racial and ethnic differences in the distribution of income and purchasing power would certainly be expected to account for at least part of the difference in suburban patterns, so long as housing costs differ among jurisdictions.

Various applications of the technique of "indirect standardization," however, indicate economic differences between majorities and minorities and account for only a small component of differences between groups in residential pattern or rate of suburbanization (Simkus, 1978). This approach compares the actual difference between majority and minority residential patterns in the aggregate against the degree of difference that would arise if the minority population in each class of income were distributed in the same way as the majority in each subarea—tract or jurisdiction—of the region. Dissimilarity, exposure (Lieberson & Carter, 1982), or other similar indices can then be determined with and without indirect standardization and compared.

TABLE 5.2
Black Residential Segregation in
Central Cities and Incorporated Suburbs,
Selected Metropolitan Areas, by Region, 1970 and 1980

Region and City (CMSA Rank in 1980 in Parentheses)	*Central City*			*Suburbs*		
	Percentage Black, 1980	*Index of Dissimilarity*[1] *1970*	*1980*	*Percentage Black, 1980*	*Index of Dissimilarity*[2] *1970*	*1980*
Northeast						
New York (1)	25	77	75	13	48	49
Philadelphia (4)	38	84	88	15	61	62
Boston (7)	22	84	80	3	42	45
Pittsburgh (13)	24	86	83	7	52	55
Newark[3]	58	76	76	22	54	60
North Central						
Chicago (3)	40	93	92	6	77	70
Detroit (6)	63	82	73	6	87	80
Cleveland (11)	44	90	91	8	77	73
St. Louis (14)	46	90	90	17	73	73
Cincinnati (20)	34	84	79	6	67	63
South						
Washington, D.C. (8)	70	79	79	15	52	53
Miami (12)	25	NA	NA	6	63	64
Baltimore (15)	55	89	86	9	NA	NA
Atlanta (16)	67	92	86	19	43	51
New Orleans (27)	55	84	76	13	NA	NA
West						
Los Angeles (2)	17	90	81	7	74	66
San Francisco (5)	13	NA	NA	7	66	53
Houston (9)	28	93	81	5	60	47
Dallas (10)	29	96	83	5	53	39
Denver (21)	12	84	70	2	40	49

SOURCES: City data are from Taeuber, 1983; Suburban data are from Logan & Schneider, 1984, Table 1.
1. Based on census block data.
2. Reflects segregation between, not within, incorporated suburban municipalities.
3. Part of N.Y. CMSA.

Comparison reveals to what extent intergroup differences in income distribution account for differences in residential pattern. Virtually all studies using this approach find that economic differences between blacks and whites, based on income, or housing cost, or occupation, account for only a relatively small part—as little as 20%—of the total level of segregation, whether among tracts in cities and SMSAs (Schnare's contribution to DeLeeuw, Schnare, & Struyk, 1976; Taeuber & Taeuber, 1965), or among suburban jurisdictions (Detroit: Hermalin & Farley, 1973; Long Island: Logan & Stearns, 1981), or between cities and suburbs (Clark in Lake, 1981). Most contribute to the impression that the fraction of total actual segregation in metropolitan areas arising from mere socioeconomic differences has been in decline since the onset of massive middle-class suburbanization after World War II (see also Farley, 1979).

Little documentation is yet available on racial or ethnic segregation within suburban rings since the mid-1970s (see Johns & Obermanns, 1985; Marshall & Stahura, 1979). But black-white dissimilarity indices have been assembled for both the cities and suburban rings of the 20 metropolitan areas already identified for 1970 and 1980. Among these 20, central city indices held fairly steady during the period; except in the West where significant declines were accompanied by rapid increases in the suburban selectivity of blacks moving to the SMSA and moving within the suburban ring (Table 5.2). Suburban dissimilarity indices, however, increased, but only slightly in the Northeast and South. Far more substantial *declines* occurred in most rings of metropolitan areas in the North Central region and the West. In the North Central region, the index declined an average of –5.6%, with Chicago and Detroit leading. In the West, change averaged –12%, but Denver's index actually rose 23% and, at the other extreme, Houston's fell by 28% and Dallas' by 26%.

Overall, modest improvements in central cities everywhere were coupled with larger gains in integration in the Northeast's suburbs, and larger losses in the suburbs of the North Central region. Ironically, the autonomous effect of the migration of higher-income blacks to the suburbs from cities may be a decrease in segregation in both the city and suburban ring. This would occur when their city origins are concentrated and their suburban destinations are diffuse. The migration to suburbs from cities of lower-income blacks, however, may actually increase segregation in cities in suburban rings, at least when their origins are diffuse and their destinations are concentrated. Simultaneous change in black-white segregation in central cities and suburban rings during the period 1970-1980 is summarized in Figure 5.2. Larger gains in central city integration occurred in the South and especially in the West. But as in the Northeast, central city gains were joined to losses in suburban integration in the South. The West emerges as the most favorable environment for integration. Here the average central city gain in black-

ABSOLUTE CHANGE IN DISSIMILARITY INDEX, SUBURBAN RING: SEGREGATION IS--	ABSOLUTE CHANGE IN DISSIMILARITY INDEX, CENTRAL CITY: SEGREGATION IS -- DECLINING	STABLE	INCREASING
DECLINING	CINCINNATI DALLAS DETROIT HOUSTON LOS ANGELES	CHICAGO CLEVELAND	
STABLE		NEW YORK ST. LOUIS WASHINGTON, D.C.	PHILADELPHIA
INCREASING	ATLANTA BOSTON DENVER PITTSBURGH	NEWARK	

SOURCE: Based on data in Table 5.2.
NOTES: *Stable* denotes absolute change between minus and plus 2 in the index. *Decline* is less than −2, and *Increase* is more than +2.

Figure 5.2 Change in Racial Segregation in Central Cities by Change in Segregation in Suburban Rings, 1970-1980

white integration was 14%, and the average gain in the suburbs was 12%. Denver's anomalous loss in suburban integration during 1970 through 1980 is paired against the largest improvement in central city integration in the West.

CONCLUSIONS

In the last two decades, minority suburbanization has accelerated steadily, and the process has broadened to include not only blacks but also Hispanics and certainly others. Black suburban selectivity rates have increased for movers whose initial addresses were in the central city or outside the metropolitan areas. Hispanic counterparts uniformly exceed the white. Significant welfare gains for the individual migrant seem assured only if suburbanward migrants secure entry to the more favored suburban jurisdictions. These fall outside the inner ring of older communities surrounding the central city and the free-standing minority enclaves scattered around the suburban ring, both prime black destinations in the past. They offer superior public services, lower tax rates,

(+/o/- show direction
of change in index, 1970–80)

PART A

BLACK-WHITE SEGREGATION IN 1970	BLACK-WHITE SEGREGATION IN 1980: LOW	BLACK-WHITE SEGREGATION IN 1980: HIGH
LOW	ATLANTA+ BOSTON+ DALLAS- DENVER+ NEW YORK- PITTSBURGH+ WASHINGTON, D.C.+	NEWARK+
HIGH	SAN FRANCISCO- HOUSTON-	CHICAGO- CINCINNATI- DETROIT- LOS ANGELES- MIAMI+ PHILADELPHIA+ ST. LOUIS°

PART B

PERCENT BLACK IN SUBURBS, 1980[3]	SEGREGATION IN 1980: LOW	SEGREGATION IN 1980: HIGH
LOW	BOSTON+ DALLAS- DENVER+ HOUSTON- PITTSBURGH+ SAN FRANCISCO-	CHICAGO- CINCINNATI- CLEVELAND- DETROIT- LOS ANGELES- MIAMI+
HIGH	ATLANTA+ NEW YORK- WASHINGTON, D.C.+	NEWARK+ PHILADELPHIA+ ST. LOUIS°

SOURCE: Based on data in Logan & Schneider, 1984, Table 1.
NOTES: 1. Partitioned at the mean dissimilarity index for all 18 cases, where "low" is less than 58.4; 2. Partitioned at the mean index, where "low" is less than 60.5; 3. Low denotes less than the mean of 9.6%.

Figure 5.3 Racial Segregation in Suburban Rings in 1980 by Suburban Racial Composition in 1980, and by Suburban Segregation in 1970

better job access, and a stronger prospect for accumulating wealth through home ownership (Lake, 1981; Logan & Schneider, 1984).

Minority welfare gain is the sum of the gains realized by individual migrants moving to the suburbs from nonsuburban locations less whatever losses may accrue to those left behind. Such losses would likely be directly proportionate to the wealth of those leaving, but inverse to the degree of integration in the region of origin. The degree of impact on the minority population is dependent on both the magnitude of the net flow of migrants to the suburban ring and the characteristics of migrants themselves as well as their origins and destinations.

During the 1970s, most suburbanward blacks moved from cities having high segregation to suburban rings characterized by somewhat less segregation. Of the 20 metropolitan regions examined, only in Detroit did the degree of the suburbs' segregation exceed the city's. Over the decade, however, accelerating black suburbanization produced little change in suburban segregation in 15 of the 18 areas for which data were available (Figure 5.3, Part A). In Newark, suburban segregation progressed from a relatively low to a relatively high level. The trend was the opposite in San Francisco and Houston. A strong correlation between percentage black and segregation in suburbs had not emerged by 1980, however (Figure 5.3, Part B). The most positive pattern—high percentage black, and low segregation—appears only in three cases: Atlanta, New York, and Washington, D.C. The least favorable condition is exhibited in Chicago, Cincinnati, Cleveland, Detroit, Los Angeles, and Miami. In these, only a very low percentage of suburban residents are black, and suburban segregation is high.

From these diverse pieces of evidence it is possible to distill a tempered but hopeful essence. Black and Hispanic households alike have almost certainly realized progress through migration to suburban communities during the last two decades. But though the pace of minority suburbanization and the rate of suburban increase have been high, the absolute number of persons affected is small. Suburbanward migration and the lessening of suburban segregation are positive signs of individual advance. Still, the larger problem of central city poverty persists and is not likely to be substantially diminished in the foreseeable future through out-migration.

REFERENCES

Clark, T. (1979). *Blacks in suburbs: A national perspective.* New Brunswick, NJ: Rutgers University, Center for Urban Policy Research.

Clark, T. (1981). Race, class and suburban housing discrimination: Alternate judicial standards of proof and relief. *Urban Geography, 2*(4), 327-338.

Clark, T. (1982). Federal initiatives promoting the dispersal of low-income housing in suburbs. *Professional Geographer, 34*(2), 136-146.

Clark, T. (1984). Suburban economic integration: External initiatives and community responses. In D. Herbert & R. J. Johnston (Eds.), *Geography and the urban environment* (Vol. 6) (pp. 213-244). New York: John Wiley.

Deleeuw, F., Schnare, A., & Struyk, R. (1976). Housing. In W. Gorham & N. Glazer (Eds.), *The urban predicament* (pp. 119-178). Washington, DC: Urban Institute.

Elgie, R. (1979). The segregation of socioeconomic groups in urban areas: A comment. *Urban Studies, 16*, 191-195.

Farley, R. (1979). *Can blacks afford to live in white residential areas? A test of the hypothesis that subjective economic variables account for racial residential resegregation.* Ann Arbor: University of Michigan, Population Studies Center.

Frey, W. (1978). Black movement to the suburbs: Potentials and prospects for metropolitan wide integration. In F. D. Bean & W. P. Frisbie (Eds.), *The demography of racial and ethnic groups* (pp. 79-117). New York: Academic Press.

Frey, W. (1984, December). Life course migration of metropolitan whites and blacks and the structure of demographic change in large central cities. *American Sociological Review, 49*, 803-827.

Frey, W. (1985, May). Mover destination selectivity and the changing suburbanization of metropolitan whites and blacks. *Demography, 22*(2), 223-243.

Goodman, J., & Streitwieser, M. (1983). Explaining racial differences: A study of city to suburb residential mobility. *Urban Affairs Quarterly, 18*(3), 301-325.

Grier, G., & Grier, E. (1983). *Black suburbanization in the 1970's: An analysis of census results.* Washington, DC: Grier Partnership.

Guest, A. (1978, December). The changing racial composition of suburbs, 1950-1970. *Urban Affairs Quarterly, 14*, 195-206.

Hermalin, A., & Farley, R. (1973, October). The potential for residential integration in cities and suburbs: Implications for the busing controversy. *American Sociological Review, 38*(5), 595-610.

Jackson, K. (1985). *Crabgrass frontier: The suburbanization of the United States.* New York: Oxford University Press.

James, F., & Crow, E. (1984). *Stronger administrative enforcement of federal fair housing laws? Lessons from the states.* Denver: University of Colorado, Institute for Urban and Public Policy Research.

James, F., Crow, E., McCummings, B., & Tynan, E. (1984). *Minorities in the sunbelt: Neighborhood segregation and housing discrimination in Denver, Houston and Phoenix.* New Brunswick, NJ: Rutgers University, Center for Urban Policy Research.

Johns, J., & Obermanns, R. (1985). *A report on population and race: Estimates of racial composition for Cuyahoga and select outlying communities 1978-1983.* Cleveland, OH: The Cuyahoga Plan.

Lake, R. (1981a). *The new suburbanites: Race and housing in the suburbs.* New Brunswick, NJ: Rutgers University, Center for Urban Policy Research.

Lake, R. (1981b, January). The fair housing act in a discriminatory market. *Journal of the American Planning Association, 47*, 48-58.

Lake, R. (1984, Spring). Changing symptoms, constant causes: Recent evolution of fair housing in the United States. *New Community, 11*, 206-213.

Lieberson, S. (1981). An asymmetrical approach to segregation. In C. Peach, V. Robinson, & S. Smith (Eds.), *Ethnic segregation in cities* (pp. 61-82). London: Croom Helm.

Lieberson, S., & Carter, D. (1982, September). Temporal changes and urban differences in residential segregation: A reconsideration. *American Journal of Sociology, 88*, 296-310.

Logan, J. (1978, September). Growth, politics and the stratification of places. *American Journal of Sociology, 84*(2), 404-416.

Logan, J., & Schneider, M. (1984, January). Racial segregation and racial change in American suburbs, 1970-1980. *American Journal of Sociology, 89*(4), 874-888.

Logan, J., & Stearns, L. (1981, September). Suburban racial segregation as a non-ecological process. *Social Forces, 60*(1), 61-73.

Long, L., & DeAre, D. (1981, September). The suburbanization of blacks. *American Demographics, 3*, 17-21.

Marshall, H., & Stahura, J. (1979, Spring). Determinants of black suburbanization: Regional and suburban size category patterns. *Sociological Quarterly, 20*, 237-253.

Nelson, K. (1979). *Recent suburbanization of blacks: How much, who and where?* Washington, DC: Department of Housing and Urban Development.

Rose, H. (1976). *Black suburbanization.* Cambridge, MA: Ballinger.

Schneider, M., & Logan, J. (1982, December). Suburban racial segregation and black access to local public resources. *Social Science Quarterly, 63*, 762-770.

Schnore, L. (1976). Black suburbanization, 1930-70. In B. Schwartz (Ed.), *The changing face of the suburbs.* Chicago: University of Chicago Press.

Schwartz, B. (1980, September). The suburban landscape: New variations on an old theme. *Contemporary Sociology, 9*, 640-650.

Simkus, A. (1978, February). Residential segregation by occupation and race in ten urbanized areas, 1950-1970. *American Sociological Review, 43*, 81-93.

Sorensen, A., Taeuber, K., & Hollingsworth, L., Jr., (1975, April). Indexes of racial residential segregation for 109 cities in the United States, 1940 to 1970. *Sociological Focus, 8*, 125-142.

Spain, D., & Long, L. (1981). *Black movers to the suburbs.* Washington, DC: Bureau of the Census.

Taeuber, K. (1983). *Racial residential segregation, 28 cities, 1970-1980* (Working Paper 83-12). Madison: University of Wisconsin, Center for Demography and Ecology.

Taeuber, K., & Taeuber, A. (1965). *Negroes in cities.* Chicago, IL: Aldine.

U.S. Bureau of the Census. (1973). *Census of population: 1970* (Subject report PC(2)-2C). Washington, DC: Government Printing Office.

U.S. Bureau of the Census. (1981, 2, 3). *Annual Housing Survey* (SMSA reports Part D). Washington, DC: Government Printing Office.

U.S. Bureau of the Census. (1981, 3, 3, and 4 respectively). *Current Population Reports*, Series P-20, Numbers 368, 377, 384 and 393. Washington, D.C.: Government Printing Office.

U.S. Bureau of the Census. (1984). *Census of population: 1980* (Subject report PC(80)-2-26) Washington, D.C.: Government Printing Office.

Wienk, R., Reid, C., Simonson, J., & Eggers, F. (1979). *Measuring racial discrimination in American housing markets: The housing market practices survey.* Washington, DC: Department of Housing and Urban Development.

6

The New Segregation: Asians and Hispanics

LOUIE ALBERT WOOLBRIGHT
DAVID J. HARTMANN

SINCE WORLD WAR II, Asia and Latin America have replaced Europe as the major sources of immigration to the United States. This change is having marked effects on American society, changing the racial and ethnic composition of the population and having important implications for many major metropolitan areas. Like past immigrants, these newcomers are placing increasing demands on major urban social institutions at a time when cities are facing severe fiscal strains and are in a period of decay. For example, schools are faced with an influx of many students who speak little English and are poorly prepared to cope with the work that they are expected to perform. In addition, these groups are often segregated in certain areas of cities and their suburbs. In many ways, these new waves of immigrants will face different problems and will have to make unique adjustments to life in modern day America.

The tremendous increase in the size of some of the newly emergent groups is striking, as can be seen in Table 6.1. Over five times as many Koreans, for example, were in the United States in 1980 as in 1970, while the Filipino population more than doubled. Four million more Mexicans were in the United States in 1980 than in 1970, an increase of almost 100% in a decade. The Cuban increase has been even larger than

AUTHORS' NOTE: *This chapter represents the opinions and views of the authors and not those of the Department of Health and Human Services or the National Center for Health Statistics.*

TABLE 6.1
Population of Various Ethnic Groups
and Increase 1970-1980

Group	*Population*		*Percentage Increase*
	1970	*1980*	
Japanese	591290	700974	18.5
Chinese	435062	806040	85.3
Filipino	343060	774652	125.8
Puerto Ricans	1429396	2013945	40.9
Mexicans	4532435	8740439	92.8
Cubans	544600	803226	47.5
Koreans	69510	354593	410.1
Vietnamese	–	261729	–

SOURCE: U.S. Bureau of the Census, 1984b: Table 52.

appears in Table 6.1. Because almost 125,000 Cubans arrived in the Mariel boat lift between April 21 and October 10, 1980, they were not counted in the 1980 census. With these additional immigrants, the Cuban population in October of 1980 was almost 930,000 for an increase of more than 70% over a decade. Southeast Asians also constitute a rapidly growing group. In 1970, there were very few Vietnamese in the United States, for example. But over 260,000 were counted in the 1980 census.

This trend is even more evident when one looks at the percentage of the foreign born population of the different groups arriving between 1975 and 1980 in Table 6.2. For example, over 90% of foreign born Vietnamese came to the U.S. between 1975 and 1980 as well as 72% of Iranians. Countries with major social and political turmoil were sending large numbers of refugees to America. Between 1975 and 1984, more than 715,000 Indochinese refugees entered the United States, including 462,000 Vietnamese, 110,000 Cambodians, 89,000 Laotians, and 55,000 Hmong (Rumbaut, 1985, pp. 433-434). Since these were the first large numbers of immigrants from these countries, there were no preexisting communities of Indochinese to help them adjust to American society.

In addition, many Central Americans also fled the political violence and social disruptions in many countries and arrived in the United States in large numbers. These trends continue. More than half of the foreign born Salvadorians, for example, arrived between 1975 and 1980. Many have been assisted by the growing sanctuary movement in the United States.

TABLE 6.2
Percentage of Foreign Born Arriving 1975-1980
for Selected Groups

Place of Birth	*Percentage*
Europe	23.7
Asia	47.0
China	27.2
Hong Kong	34.9
India	43.7
Iran	71.9
Japan	31.6
Korea	52.3
Philippines	34.4
Vietnam	90.5
El Salvador	51.3
Mexico	33.0
Cuba	6.3
South America	32.6
Africa	43.9

SOURCE: U.S. Census Bureau, 1985: Table 87.

In total, 47% of Asian immigrants arrived between 1975 and 1980 (see Table 6.3). More than a million Asian immigrants entered the United States between 1981 and 1984, compared to 260,000 from Europe. Significantly, more immigrants, 274,000, arrived from Mexico than all of Europe. China, Korea, and the Philippines also had more than 100,000 immigrants. Vietnam, Cambodia, and Laos supplied over 350,000 of the Asian immigrants. Southeast Asia will continue to provide a large share of refugees since large numbers of people are waiting in refugee camps to come to America.

THE REGIONAL DISTRIBUTION OF THE NEW GROUPS

Many of the new ethnic/racial groups are segregated, both by region and by their location within metropolitan areas. Regional differences occur in terms of economic growth, employment levels, income, educational and occupational distributions, and other factors. Therefore, where immigrant groups settle may have a significant effect on the nature and rate of their assimilation into American society. As can be seen in Table 6.4, the white population is distributed evenly throughout

TABLE 6.3
Number of Immigrants Arriving 1981-1984 in Thousands

Place of Origin	*Number*
Europe	258.8
Asia	1111.6
China	141.1
India	93.7
Japan	15.9
Korea	130.7
Philippines	173.2
Vietnam	202.9
Cambodia[a]	62.2
Laos[a]	99.4
Mexico	274.1
Cuba	38.6

SOURCE: U.S. Bureau of the Census, 1985: Tables 86-87.
a. Includes only those admitted under refugee acts.

the nation's regions. No other group is so distributed. The significant movement of the white population to the Sunbelt is clearly evident. Indeed, the South now has the highest proportion of whites, surpassing the North Central region in the 1970s.

The new immigrant groups are heavily concentrated in the West. The Japanese are the most heavily concentrated group, with over 80% of them in the West in both 1970 and 1980. Nevertheless, for the older Asian groups, the Japanese, Chinese, and Filipinos, there has been a slight movement out of the West, though over 50% of each of these groups remain there.

The Hispanic groups, with the exception of Cubans, have joined whites and blacks in moving to the West, as the proportion of persons of Spanish origin there rose from 37% to 43% between 1970 and 1980. Puerto Ricans and Cubans are less concentrated in the Northeast, with the proportion of Cubans declining by almost 10% and that of Puerto Ricans by 8% in the 1970s. Cubans have become increasingly concentrated in the South, with 60% in that region, while 90% of Mexican immigrants were in the South and West in both 1970 and 1980.

California has become the new Ellis Island, especially the Los Angeles area. One of every four foreign born persons lives in California, with one of every eight in the Los Angeles SMSA. Table 6.5 shows that between 1970 and 1980, with the exception of Cubans, all foreign born groups found a greater proportion of their population in California.

TABLE 6.4
Region of Residence of Various Groups
1970 and 1980 in percentages

Group	1980 NE	1980 NC	1980 S	1980 W	1970 NE	1970 NC	1970 S	1970 W
White	22.5	27.7	31.3	18.5	24.9	29.1	28.4	17.7
Black	18.3	20.2	53.0	8.5	19.2	20.2	53.0	7.5
Spanish Origin	17.8	8.7	30.6	42.8	20.9	11.6	30.4	37.1
Mexicans	1.0	9.3	35.5	54.1	1.0	8.3	37.5	53.3
Cubans	22.8	4.1	63.9	9.2	32.2	6.0	51.9	9.9
Puerto Ricans	73.8	10.4	9.1	6.8	81.3	9.4	4.5	4.8
Japanese	6.7	6.3	6.4	80.6	6.6	7.2	51.	81.0
Chinese	27.0	9.1	11.2	52.7	26.6	9.0	7.9	56.5
Filipino	9.7	10.3	10.7	69.3	9.2	8.1	9.3	73.4
Koreans	19.2	17.5	19.9	43.4	–	–	–	–
Vietnamese	9.5	14.0	30.7	45.8	–	–	–	–

SOURCE: U.S. Bureau of the Census, 1973a: Table 1; 1973c: Table 1; 1983b: Tables 50, 51, 52, 233.

Almost a third or more of each of the Asian groups now resides in California. In addition, one of every five Mexicans lives in Los Angeles County and at least 10% of each of the Asian groups lives there. No other city or county comes close to this concentration of Asians and Hispanics, including New York. Substantial Cuban, Chinese, and Puerto Rican populations do reside in New York. Nevertheless, Puerto Ricans are less concentrated in the New York SMSA: almost 60% lived there in 1970 compared to 44% in 1980.

THE DISTRIBUTION OF ASIANS AND HISPANICS WITHIN METROPOLITAN AREAS

During the 1970s, whites left metropolitan areas in significant numbers for rural areas and small towns. The proportion of the white population in the suburbs fell by 6%, while the central city proportion declined by 3%. The black population had a slight decline in its concentration in central cities and a major decline in its rural component, while the black suburban population rose significantly, although in concentrated areas.

Unlike the white population, the Asian population is almost completely centered in metropolitan areas, with less than 10% in rural areas or small towns. For all groups shown in Table 6.6, there was a remarkable rate of suburbanization and a significant decline in the

TABLE 6.5
Percentage of Groups Living in Selected Areas,
1970 and 1980

	California		*Los Angeles SMSA*		*New York SMSA*	
Group	*1970*	*1980*	*1970*	*1980*	*1970*	*1980*
Japanese	36.1	37.4	17.6	16.6	2.9	3.4
Chinese	39.1	40.0	9.4	11.6	17.5	16.6
Mexicans	41.0	41.6	18.1	18.9	–	0.3
Cubans	8.7	7.6	7.2	5.5	7.5	8.9
Puerto Ricans	3.6	4.6	1.6	1.8	58.8	44.3
Filipinos	40.2	46.1	9.5	12.8	3.7	3.9
Koreans	22.9	29.3	12.7	17.1	7.1	8.0
Vietnamese	–	34.2	–	11.0	–	1.5

SOURCE: U.S. Bureau of the Census, 1973a: Tables 1, 14, 60, 67; 1983b: Table 232.

central city proportions between 1970 and 1980. Only the Chinese have a concentration of population in central cities comparable to blacks. Even the newest Asian group, the Vietnamese, have a proportion in the suburbs that is twice as high as blacks, exceeding even the proportion of whites in the suburbs. Over half of the Korean population is in the suburbs, despite the fact that most have arrived in the last ten years.

The Spanish origin population is more concentrated in rural areas and central cities than Asians. However, the three Spanish origin groups differ greatly in the distribution of their populations. Cubans are the most metropolitan and have the highest proportion in the suburbs, while Puerto Ricans are most concentrated in central cities. Mexicans, on the other hand, are more concentrated in rural areas, with one of every eight Mexicans located there. Much of the urbanization of Mexican Americans has occurred in recent years. Interestingly, the Spanish origin population was less metropolitan and rural in 1980 than in 1970, indicating a movement to nonmetropolitan urban areas. For all groups, both Asian and Hispanic, there is less of a concentration of population in central cities than was true of Southern and Eastern European groups arriving between 1880 and 1920.

Table 6.7 contains indexes of dissimilarity for seven SMSAs for the segregation of whites, blacks, Hispanics, and Asians in 1980. Blacks are, by far, the most segregated group from each of the others. The lowest levels of segregation are between Asians and whites, except in San Diego, where Asians and Hispanics are less segregated from each other than either group is from whites. The rates of segregation are between 10 and 40 points higher for Asians and Hispanics with blacks than with

TABLE 6.6
Distribution of Groups in Central Cities, Suburbs, and Rural Areas, 1970 and 1980

	1970			*1980*		
Group	*Central Cities*	*Suburbs*	*Rural*	*Central Cities*	*Suburbs*	*Rural*
White	27.8	40.0	27.6	24.6	33.8	28.7
Black	58.2	16.1	18.7	57.2	18.7	14.7
Spanish Origin	49.9	32.9	12.8	48.8	30.2	10.1
Asian and Pacific Islander	–	–	–	46.0	39.9	6.8
Japanese	47.8	37.7	11.4	42.7	46.0	8.3
Chinese	67.8	25.4	3.7	58.1	38.0	3.0
Filipino	47.9	36.6	14.7	45.0	46.0	7.6
Korean	–	–	–	41.4	51.1	7.0
Vietnamese	–	–	–	52.0	40.4	4.6
Mexican	–	–	14.5	46.2	38.7	12.4
Cuban	–	–	1.5	40.2	57.1	2.2
Puerto Rican	–	–	2.3	75.3	21.5	3.0

SOURCE: U.S. Bureau of the Census, 1973a: Table 124; 1983b: Tables 50-51.

whites. On average, Asians are 27 points more segregated from blacks than whites in the six Sunbelt cities, and over 40 points less segregated in Chicago. In general, the groups are only slightly less segregated from blacks than are whites, as all groups appear to shun contact with blacks. Whites are 22 points more segregated from blacks than Hispanics in the six Sunbelt cities, almost 30 points more segregated in Chicago, and over 30 points more segregated from blacks than Asians in the seven cities. This has historically been true as shown by Grebler, Moore, and Guzman (1970), by Taeuber and Taeuber (1965), and by Massey (1979).

Significantly, segregation has been decreasing for whites and Hispanics between 1960 and 1980. The decline was by 6.9 points in the 5 Southwestern cities, but by less than two points for Hispanics and blacks. Segregation changed little in Los Angeles and San Diego, but declined greatly in Houston, by almost 17 points, followed by Denver—12 points, and Phoenix—7 points. Interestingly, segregation between blacks and Hispanics declined by 17 points in Phoenix, while remaining virtually unchanged in the other cities.

Guest and Weed (1976) compared segregation patterns for several ethnic groups for the period 1930 to 1970. They observed major decreases in segregation for Asian groups. This was not true of Mexicans and Puerto Ricans.

TABLE 6.7
Indexes of Dissimilarity for Whites, Blacks, Asians, and Hispanics in 7 SMSAs 1960 and 1980

SMSA Group	*1960*	*1980* *Blacks*	*1980* *Hispanics*	*1980* *Asians*
San Diego				
Whites	43.6	64.0	41.5	45.5
Blacks	55.2	–	50.5	50.3
Hispanics	–	–	–	38.3
Phoenix				
Whites	57.8	62.5	51.0	35.7
Blacks	60.7	–	44.3	59.1
Hispanics	–	–	–	51.9
Miami				
Whites	–	79.1	52.6	33.7
Blacks	–	–	79.9	75.8
Hispanics	–	–	–	52.3
Los Angeles				
Whites	57.4	81.2	57.2	46.7
Blacks	75.7	–	72.7	76.2
Hispanics	–	–	–	49.1
Houston				
Whites	65.2	75.0	48.6	45.4
Blacks	70.9	–	68.3	76.9
Hispanics	–	–	–	60.7
Denver				
Whites	60.0	69.7	48.2	33.9
Blacks	68.0	–	71.3	63.5
Hispanics	–	–	–	46.3
Chicago				
Whites	–	87.9	63.6	47.8
Blacks	–	–	84.0	88.9
Hispanics	–	–	–	64.5

SOURCE: 1960 from Grebler, Moore, & Guzman, 1970: p. 275. 1980 from STF-3.
NOTE: For 1960 Hispanics are persons of Spanish surname, in 1980 they are persons of Spanish origin.

FACTORS AFFECTING SEGREGATION: THE EXPERIENCE OF ASIAN AND HISPANIC GROUPS

Many factors have been noted that have affected the integration of immigrant and racial groups. One of the most fundamental is the physical appearance of persons belonging to the group. As Lieberson

(1980) notes, ethnicity is often a matter of choice for whites. Once white ethnics learned English and American customs, they could disappear into the larger society and be virtually indistinguishable from the majority. With nonwhites, this was not possible, because their skin color, hair texture, and stature made them easily recognizable. Acquired characteristics can be changed; inherited ones cannot. For those interested in perpetuating a system of dominance, physical characteristics provide easy means of identifying people for discrimination. This has occurred repeatedly in American history, with blacks, American Indians, Hispanics, and Asians.

Each of the Asian and Hispanic groups has had a different pattern of assimilation, which is reflected in the social, economic, and housing tenure characteristics shown in Tables 6.8, 6.9, 6.10. In general, the Asian groups are of much higher social status than the Hispanic groups; members are older, better educated, with higher incomes, smaller proportions of families below the poverty level, and higher rates of home ownership.

MEXICAN AMERICANS

Greater differences exist between different segments of the Mexican American population than any of the other Asian and Hispanic groups. Mexican Americans are found throughout the class structure and in almost every neighborhood in the Southwest. Members differ greatly in social class and time of arrival, ranging from super rich to oppressively poor, from descendants of sixteenth-century Spanish settlers to illegal immigrants newly arrived from across the border. Some Hispanics are indistinguishable from other whites, some are quite dark. Generally, complexion has been highly related to assimilation and access to housing, with those of light complexion being most easily absorbed into the mainstream of society.

Despite some success in assimilation, Mexican Americans have suffered from a history of discrimination, subordination, and segregation. Many entered the United States as a conquered people and quickly lost their lands and became disenfranchised. Discrimination denied them education and employment skills, and most occupy the lowest occupational status in the Southwest. Also, since Mexican Americans are the group with the highest proportion of illegal residents, many are exploited and abused, particularly in the areas of employment and housing.

TABLE 6.8
Selected Social Characteristics by Group, 1970 and 1980

	Median Age		*Percent Female Headed Families*		*Median Years Schooling*		*Percentage High School Graduates*		*Percentage Foreign Born*	
Group	*1970*	*1980*	*1970*	*1980*	*1970*	*1980*	*1970*	*1980*	*1970*	*1980*
Whites	28.9	31.3	9.0	11.2	12.1	12.5	54.6	68.8	4.9	3.9
Blacks	22.5	24.9	27.4	37.8	9.8	13.0	31.4	51.2	1.1	2.8
Spanish Origin	21.1	23.2	15.3	19.9	9.1	10.8	32.1	44.0	15.6	28.6
Japanese	32.3	33.6	10.3	12.3	12.5	12.9	68.8	81.6	20.9	28.4
Chinese	26.8	29.6	6.7	8.6	12.4	13.4	57.8	71.3	47.1	63.3
Filipino	26.2	28.6	8.6	12.1	12.2	14.1	54.7	74.2	53.1	64.7
Korean	26.4	26.1	14.7	11.5	12.9	13.0	71.1	78.1	54.0	81.9
Vietnamese	–	21.2	–	14.8	–	12.4	–	62.2	–	90.5
Mexican	19.3	21.9	13.4	16.4	8.1	9.6	24.2	37.6	18.0	26.0
Puerto Ricans	20.3	22.3	24.1	35.3	8.6	10.5	23.4	40.1	–	3.0
Cubans	31.7	37.7	12.3	14.9	10.3	12.2	43.9	55.3	81.6	77.9

SOURCE: U.S. Bureau of the Census, 1983b: Tables 43, 47, 48, 143,160, 166, 161, 167; 1973a: Tables 190, 207, 189, 215.
NOTE: Median years of schooling and percentage high school grads based on population 25 years of age and older.

TABLE 6.9
Selected Economic Characteristics by Group, 1970 and 1980

Group	*Median Family Income* 1970	1980	*Percentage of Families Below Poverty Level* 1970	1980	*Unemployment Rate* 1970	1980
Whites	9958	21014	8.6	7.0	4.1	5.7
Blacks	6063	12627	29.9	26.5	7.0	11.8
Spanish Origin	7348	14712	21.2	21.3	6.4	8.9
Japanese	12515	27354	8.4	4.2	2.0	3.3
Chinese	10610	22559	10.3	10.5	3.0	4.1
Filipinos	9318	23687	11.5	6.2	4.7	4.6
Koreans	–	20459	–	13.1	3.6	6.9
Vietnamese	–	12840	–	35.1	–	8.2
Mexicans	6962	14765	24.4	20.6	7.1	9.1
Puerto Ricans	6115	10734	27.1	34.9	6.2	11.7
Cubans	8529	18245	13.3	11.7	5.6	6.0

SOURCE: U.S. Bureau of the Census, 1973a: Tables 215, 250; 1983b: Tables 149, 162, 164, 165, 168, 170, 171.
NOTE: Unemployment rate is based on population 16 years of age and older.

Despite efforts to prevent illegal immigration, the twentieth century has witnessed three major waves of Mexican immigration. The first was composed of refugees fleeing the violence of the Mexican Revolution. Large numbers of Mexicans were uprooted by the decade of violence and turmoil from 1910 to 1920 and found their way to the United States. This first major stream of migration ended with the Great Depression, when many Mexicans were deported to Mexico. Discrimination against Mexican Americans increased during this period.

The second wave of immigration resulted from the demand for labor during World War II. Thousands of Mexicans came north to work in war-related industries. With the difficult economic times following the war, efforts were renewed to curtail illegal immigration and to expel those living in this country illegally. Many Mexican Americans were driven from their jobs by returning veterans.

Finally, the last wave of Mexican immigration began in the 1960s as a result of the Vietnam War boom. This third wave continues today and has changed dramatically the composition of the Mexican population. The virtual collapse of the Mexican economy in the 1980s continues to push immigrants northward.

This continued migration has hindered the assimilation of the Mexican Americans. Segregation is not greatly reduced by the move-

TABLE 6.10
Percentage of Owner Occupied Housing Units
by Group, 1970 and 1980

Group	*1970*	*1980*
Whites	65.4	67.8
Blacks	41.5	44.4
Spanish Origin	43.7	43.4
Japanese	56.1	59.0
Chinese	43.8	54.5
Filipino	39.7	55.9
Puerto Ricans	15.1	20.7
Mexicans	51.2	48.9
Cubans	27.8	43.9

SOURCE: U.S. Bureau of the Census, 1973b: Tables 10, 25, 40; 1973c: Table 12; 1984a: Tables 6, 7, 8, 9.

ment of Mexican Americans out of the barrios, because they are replaced by newcomers from Mexico. Thus overall levels of segregation mask significant differences between the native born and the foreign born and between different generations of the native born (Massey, 1981). The newcomers have little anticipatory socialization and bring little human capital with them, reducing the likelihood that they will be able to advance in the modern American economy. They are prepared only for jobs requiring few skills and little education. In addition, many come as sojourners, with little intention of remaining in the United States.

Nevertheless, Mexican American segregation is less than one would expect given their low social and economic status and the high proportion foreign born. Indeed, Mexican American segregation is comparable to that of many European groups early in this century. Also, surprisingly, Mexican Americans had a high rate of home ownership, with their rate in 1970 trailing only the whites and Japanese. During the 1970s, home ownership declined relative to Asian groups other than the Japanese.

In sum, Mexican American segregation is moderate by historical standards. Significantly, as Mexican Americans rise in social status, they are able to leave the barrios and enter neighborhoods in almost any area where they can afford housing. In Chicago and Los Angeles, they are entering areas where blacks can enter only if they are of very high status (Woolbright, 1985). In fact, some areas of the Chicago SMSA have allowed Hispanics to enter so that they could claim to be integrated, thereby justifying the exclusion of blacks.

Paradoxically, then, Mexican Americans have a high rate of acceptance into white neighborhoods, while at the same time large numbers are becoming trapped in inner-city barrios. This is especially true of the more recent immigrants, with their high fertility, low levels of educational attainment, high rates of unemployment, and high proportions of female-headed families. With the resurgence of illegal immigration, the problem of segregation is increasing and cleavages in the Mexican American population between the "haves" and "have nots" are widening.

PUERTO RICANS

Of all the Asian and Hispanic groups, Puerto Ricans are the most segregated and have the lowest socioeconomic status. Puerto Ricans are one of the newer groups in American society. Though the island became a United States possession as a result of the Spanish American War, it was only during the 1940s that large numbers of Puerto Ricans began to arrive. In 1940, there were only 70,000 Puerto Ricans in the United States. However, since that time, the population exploded to 226,000 in 1950, 615,000 in 1960, 1.4 million in 1970, and over two million in 1980. This sudden, very rapid increase makes assimilation extremely difficult.

Furthermore, there has been little time for generational benefits to develop, since the average Puerto Rican has spent less than twenty years in the United States. In addition, there is much back and forth movement between the mainland and the island, which may limit commitments and reinforces Puerto Rican cultural survivals. A network of interpersonal relations provides assistance in the United States and reinforces contacts with the island.

Several additional factors operate to make Puerto Ricans one of the most segregated groups in American society. In appearance, many resemble blacks, because of the high rate of racial intermixture on the island. As a result, many Americans consider them to be blacks, though they classify themselves as white. Second, like Mexicans, many arrive in the United States with low levels of human capital and little anticipatory socialization. Their high proportion of female-headed families, high unemployment rates, low incomes, and high percentage of families below the poverty level are close to those of blacks. Indeed, no other group so closely approximates the low socioeconomic status of blacks. All of these factors are reflected in low rates of home ownership.

Many Puerto Ricans arrived during the 1950s, when economic opportunities were relatively restricted in many Northern cities where they first settled. The central cities where Puerto Ricans settled were devastated by economic and social decay. Puerto Ricans entered urban areas with a decaying and aging housing stock, a high crime rate, social disorganization, and high levels of segregation. For many whites, these characteristics soon became associated with Puerto Ricans. Many whites began to fear that the coming of Puerto Ricans signaled neighborhood decay, crime, and social disorganization.

As a result, Puerto Ricans are as residentially segregated as blacks, having low rates of suburbanization and being confined to the inner cities of the Northeast. Puerto Ricans are, in fact, more likely to live in central cities than blacks. Prospects for change are not encouraging, given low socioeconomic status and continued white flight.

CUBANS

Cubans have an interesting socioeconomic profile, which has resulted in their having the greatest assimilation of all the Hispanic groups. They most closely match whites on all the social and economic factors in Tables 6.8 and 6.9, despite several factors that should have limited their assimilation. Cubans are very recent immigrants; only a small community of Cubans was in the United States before 1960 when thousands were uprooted by the Cuban Revolution. Thus few received assistance from friends and relatives upon arrival and there has not been time for generational effects to arise. As late as 1980, almost 80% of Cubans were foreign born.

Nevertheless, Cuban assimilation has been relatively successful, because of several advantages that Cubans have over Mexican Americans and Puerto Ricans. First, Cubans are a highly select group; though they did not have time for much anticipatory socialization, refugees were largely middle-class whites. These are the persons who could most take advantage of opportunities in America. In addition, they had no choice but to assimilate; there was no going back to the society that they left. Also, migration was seriously curtailed after the first mass arrivals, so that Cuban cultural patterns and the use of Spanish were not being constantly reinforced by new streams of immigrants.

Cubans also benefited from a number of government programs that facilitated their assimilation. The government established special programs to teach them English, to provide them with vocational training or retraining, and job placement. Also, the government sought to distribute them around the country, so that they would not be segregated in one part of the country. This enabled them to take advantage of both the national job markets and housing markets. Government programs also helped Cubans gain certification and licensing in their middle-class jobs, so that their Cuban credentials could be validated and transferable (Pedraza-Bailey, 1980). All of these programs were unavailable to Mexicans and Puerto Ricans. The importance of these government programs cannot be overemphasized.

Nevertheless, the condition of Cubans may have deteriorated in recent years. First, Cubans have become more concentrated in South Florida. The proportion living in the South rose from 52% in 1970 to 64% in 1980. As Winsberg (1979) points out, this increasing concentration in the Miami area has increased their level of segregation, though Cubans have been able to penetrate older middle-class neighborhoods throughout the SMSA.

The Mariel boat lift caused dramatic changes. One of every nine Cubans in the United States arrived at that time. These new arrivals were of much lower social status than the earlier arrivals. Many had mixed black and white ancestry. Because Castro emptied his prisons and mental institutions, these new arrivals also became associated with crime and violence. Many are still in detention centers, as the United States government vainly tries to return them to Cuba. How these Marielitos will be assimilated remains to be seen and will be an important determinant of future Cuban assimilation and segregation.

CHINESE

Chinese immigration has come in two major waves. The first influx of Chinese came in the nineteenth century and was met with intense discrimination. The first Chinese were brought to the United States as coolie laborers, performing the most difficult and arduous tasks on the railroads, the mines, and fields of the West. Wages were low and working conditions were brutal. Any economic success on the part of the Chinese was perceived as a threat to employment prospects and the working conditions of whites. Intense anti-Chinese feelings arose, which resulted in violence against the Chinese and calls for the restriction of Chinese immigration. As a result, the Chinese Exclusion Act of 1882 was passed, dramatically reducing the entry of Chinese immigrants. Since most Chinese men had left their wives and families in China, they had no way of bringing them to America. From the 1880s until the 1940s, then, the Chinese community was predominantly male. The fact that there were few Chinese women, combined with the fact that the laws in many Western states prevented Chinese from intermarrying, caused the second generation to be small.

Discrimination also extended to housing, as whites barred Chinese from living in their neighborhoods and landlords refused to rent to Chinese. As a result, the Chinese became highly segregated into Chinatowns in the major cities. Even today, the Chinese are the group most heavily segregated in central cities; not even blacks are as likely to live in central cities as are Chinese.

FILIPINOS AND KOREANS

Filipinos and Koreans are immigrating to the United States in large numbers. Almost all of them have arrived in this century. In 1920, there were only 27,000 Filipinos in the United States, but this rose to 108,000 in 1930. From then until 1960, the population rose slowly; then it exploded as the population almost doubled in the 1960s from 176,000 to 343,000. In the 1970s, it nearly doubled again to 775,000. The Korean increase has been even more dramatic, from 70,000 in 1970 to 355,000 in 1980, a 400% increase. As a result, over 80% of Koreans are foreign born, as are almost two-thirds of Filipinos.

Members of both of these groups come with high levels of skills and educational attainment. The average member of each group has had one to two years of college training. Many of these immigrants come to take advantage of shortages of trained personnel for certain types of occupations. For example, significant numbers of Filipino nurses have entered the United States to work in hospitals and nursing homes. In addition, there are strong ties among the United States and each of these countries. Both have American military bases and a relatively high degree of contact with Americans. Thus immigrants from both countries are able to obtain a degree of anticipatory socialization.

Distinctive migration patterns have also evolved. Strong family ties influence who come and where they settle. Many of the new immigrants have family and friends in the United States who help them find jobs and adjust to life in America. Both groups are concentrated in the West, but Koreans are more evenly distributed around the country than any other Asian group. Both are primarily metropolitan and have extremely high proportions in the suburbs, with Koreans being second only to Cubans. Segregation for both is low to moderate.

SOUTHEAST ASIANS

Southeast Asians consist of several groups displaced by the Vietnam conflict, including Vietnamese, Laotians, and Cambodians. This is a very heterogeneous group, varying from well-educated professionals to primitive animists. The Hmong, for example, are a primitive, rural, preindustrial people who have little education or preparation for life in the modern world. As a result, they are having great difficulty adjusting to American society. Most Southeast Asians arrived with little or no anticipatory socialization and had no community of their coethnics here to help them.

Despite the efforts of refugee programs, they have had serious adjustment problems and have shown little evidence of assimilation. This is most true of the Hmong, Laotians, and Cambodians, but applies generally to all groups.

The refugee program has attempted to scatter Southeast Asians throughout the United States. Although still relatively scattered, Southeast Asians are beginning to congregate according to clan and kinship groups, with increased migration to California. Thus segregation is increasing as refugees move to be near relatives and friends.

The Vietnamese have had more success in assimilating than the other groups, because many of them are better educated and had more contact with Americans during the war. Also, many have been in the United States longer and have had more time to adjust. As can be seen from Table 6.4, most of the Vietnamese live in the South or West. California is home to 34% of all Vietnamese, with 11% in Los Angeles County. High proportions live in the central cities and suburbs of major metropolitan areas. Because of their low incomes and lack of skills, they must find low-cost housing, which has led to relatively high levels of segregation. The median family income was only $13,000 in 1980, with 35% of Vietnamese families having incomes below the poverty level, despite the fact that the average Vietnamese had 12.4 years of schooling.

CONCLUSIONS AND POLICY RECOMMENDATIONS

Because it is felt that segregation is merely one aspect of the larger problem of assimilation, throughout this chapter the major concern has been with the adjustments of immigrants to American society and cultural norms. The newly emergent groups, with a few notable exceptions, have much more in common with earlier immigrant groups than with blacks. Thus they have a much greater chance of escaping poverty and geographic isolation than blacks, though this is far from certain for some groups. Government policies can greatly facilitate the assimilation of these groups and reduce their segregation.

Immigration laws greatly influence assimilation, for example. Immigrants to the United States fit into one of two categories: legal or illegal. Legal migrants have a much greater chance to assimilate and have lower levels of segregation. They are screened by immigration authorities and are a select group, frequently being admitted only if they are from an occupational category that is in short supply in this country or if they will be reunited with family members. As a result, they are usually of relatively high social status and have relatives and friends to assist them in assimilating. Illegal immigrants, on the other hand, have little training, few skills, and little anticipatory socialization. They come to satisfy a demand for low-skilled service and manufacturing workers at the bottom of the occupational ladder. Often, these jobs have no Americans willing to take them. As a result of their illegal status, they are exploited and live in overcrowded conditions in the slums. More open immigration laws may help with this problem.

In addition, more programs are needed to relieve urban governments of the burden placed upon them by the large influx of immigrants. The federal government must assume a greater role. This burden is especially great in many West Coast cities, which simply cannot afford the costs of assimilating the large number of Asian and Hispanic poor. More language and Head Start-type programs are needed to help immigrants and their children to learn English and American culture. The recent flood of refugees has overwhelmed the refugee assistance programs and church and community groups have not been able to meet the needs of these people. This is true of Southeast Asians, Marielitos, and Central Americans, whose situations are becoming increasingly discouraging. Since the conditions that immigrants meet on arrival often set the tone for several generations, what happens in the next few years will determine if these groups become trapped in the underclass or repeat the experience of earlier immigrant groups and are assimilated into American society. The role of the federal government is critical.

Many of these groups have had little experience with the American housing market. They must receive assistance in filling out forms, getting loans, and entering the home ownership market. When members of Hispanic and Asian groups move, they are much less likely to buy than are comparable whites. Many renters do not know how to go about buying a home. The process can be long and involved, requiring financial and legal assistance that discourage people with limited experience with American customs.

Anti-Asian prejudice may increase as these groups become more concentrated in certain neighborhoods and occupations. In addition,

Asians, especially Japanese, may be blamed for the loss of American jobs. Asians are becoming progressively more visible in many metropolitan areas and may be perceived as threats to neighborhood stability.

In conclusion, most of the newly emergent groups have shown a great propensity to assimilate into American society. Those who arrive with the highest levels of skills and anticipatory socialization have had the greatest success because of changes in the American economy and occupational structure. Those who arrive with low levels of skills, unable to speak English, and with few prospects for upward social mobility are at great risk of being trapped in a perpetual cycle of poverty. The quotas established by the immigration laws ensure that most legal immigrants will have a chance of reaching middle-class standing. The groups not covered by the regular immigration laws, the refugees, illegal aliens, and Puerto Ricans are the ones most at risk.

Segregation of Hispanic and Asians groups will continue to be moderate to relatively high, because such a high proportion of them are recent arrivals and have had little experience with American society. With time, segregation should decline as it did for earlier immigrant groups. The rate and extent, and which groups remain segregated and which do not, will depend largely on what the federal government does or fails to do.

REFERENCES

Gordon, M. (1964). *Assimilation in American life: The role of race, religion and national origins*. New York: Oxford University Press.

Grebler, L., Moore, J., & Guzman, R. (1970). *The Mexican American people: The nation's second largest minority*. New York: Free Press.

Guest, A., & Weed, J. (1976). Ethnic residential segregation: Patterns of change. *American Journal of Sociology, 81*, 1088-1111.

Jaffe, A., Cullen, R., & Boswell, T. (1976). *Spanish Americans in the U.S.: Changing demographic characteristics*. New York: Research Institute for the Study of Man.

Lieberson, S. (1980). *A piece of the pie: Blacks and white immigrants since 1880*. Berkeley: University of California Press.

Massey, D. (1979). Effects of socioeconomic factors on the residential segregation of black, and Spanish Americans in U.S. urbanized areas. *American Sociological Review, 44*, 1015-1022.

Massey, D. (1981). Hispanic residential segregation: A comparison of Mexicans, Cubans, and Puerto Ricans. *Sociology and Social Research, 65*, 311-322.

Montero, D. (1981). The Japanese Americans: Changing patterns of assimilation over three generations. *American Sociological Review, 46*, 829-839.

Pedraza-Bailey, S. (1980). *Political and economic migrants in America: Cubans and Mexicans*. Unpublished Ph.D. dissertation at the University of Chicago.

Ransford, E. (1977). *Race and class in American society: Black, Chicano and Anglo.* Cambridge, MA: Schenkman.

Rosenberg, T., & Lake, R. (1976). Toward a revised model of residential segregation and succession: Puerto Ricans in New York, 1960-1970. *American Journal of Sociology, 81*, 1142-1150.

Rumbaut, R. (1985). Mental health and the refugee experience: A comparative study of Southeast Asian refugees. In T. C. Owan (ed.), *Southeast Asian mental health: Treatment, prevention, services, training, and research* (pp. 433-486). Rockvile, MD: National Institute of Mental Health.

Taeuber, K., & Taeuber, A. (1965). *Negroes in cities: Residential segregation and neighborhood change.* Chicago: Aldine.

United States Bureau of the Census. (1973a). *1970 census of population: United States summary, Chapters A-D.* Washington, DC: Government Printing Office.

United States Bureau of the Census. (1973b). *1970 census of population: Subject reports, Japanese, Chinese and Filipinos in the U.S.* Washington, DC: Government Printing Office.

United States Bureau of the Census. (1973c). *1970 census of population: Subject reports, persons of Spanish origin.* Washington, DC: Government Printing Office.

United States Bureau of the Census. (1984a). *1980 census of housing, general housing characteristics: United States summary.* Washington, DC: Government Printing Office.

United States Bureau of the Census. (1984b). *1980 census of population: United States summary, Chapters A-D.* Washington, DC: Government Printing Office.

United States Bureau of the Census. (1985). *Statistical abstract of the United States, 1986.* Washington, DC: Government Printing Office.

Weinberg, D. (1978). Further evidence on racial discrimination in home purchase. *Land Economics, 54*, 505-513.

Winsberg, M. (1979). Housing segregation of a predominantly middle class population: Residential patterns developed by the Cuban immigration into Miami, 1950-1974. *American Journal of Economics and Sociology, 38*, 403-418.

Woodrum, E. (1981). An assessment of Japanese American assimilation, pluralism and subordination. *American Journal of Sociology, 87*, 157-169.

Woolbright, L. A. (1985). *The invasion process revisited: Hispanic, black, and Asian residential succession and transition.* Unpublished Ph.D. dissertation at the University of Chicago.

7

Market Failure and Federal Policy: Low-Income Housing in Chicago 1970-1983

JAMES W. FOSSETT
GARY ORFIELD

THE NATIONAL URBAN POLICY DEBATE in the 1980s has seen a sustained attack on the idea of major public intervention in urban housing markets, both in the provision of housing for poor people and in attacking racial barriers to equal housing opportunity. The Reagan administration has consistently argued that the private market can satisfy the housing needs of the poor far more successfully than government and has succeeded in eliminating or drastically reducing most housing programs for low-income groups. At the same time, the federal government has left intact or even expanded implicit and explicit public subsidies aimed at stimulating home ownership, primarily for middle- and upper middle-income groups. Simultaneously, the administration has opposed congressional efforts to strengthen federal fair housing laws and has redirected Justice Department activities away from broad attacks on discrimination in housing markets. Indeed, the Justice Department has been reduced to handling a small number of individual complaints.

The Reagan model, and it is not a particularly new one, relies on a combination of market forces to provide housing to the poor through a

AUTHORS' NOTE: *The research reported in this chapter was supported by the Institute of Government. We would like to acknowledge admirable research support from John A. Hamman and editorial assistance from Anna Merritt. The views and opinions in this chapter are solely those of the authors and should not be attributed to any other institution or individual.*

"trickle-down" process. At the same time, discriminatory practices will be handled only on a complaint basis, initiated by the individual who may have experienced discrimination. While the Reagan administration has made these policies explicit, they have been in practice for most of the past fifteen years.

This study attempts to test the adequacy of this "model" of federal housing policy. We examine the effectiveness of the "private" market in the city of Chicago in providing adequate, affordable, housing to lower-income and minority groups over the last fifteen years. Our conclusions are pessimistic. In spite of demographic and economic conditions that are favorable to an improvement in the housing conditions of these groups, available evidence suggests that poor and minority households are paying substantially increased shares of their income for housing that has not appreciably improved in quality. We then argue that the reason for this apparent paradox lies in the city's exceedingly segregated mechanism for allocating housing, which confines minorities, particularly blacks, into a secondary housing market that is effectively isolated from the larger, white-dominated market. We finally suggest that the consequences of Reagan administration housing policies in this type of environment will be to exacerbate the problems of minorities and the poor in locating adequate and affordable shelter.

MARKETS AND POLICIES: A REVIEW OF THE ARGUMENTS

Most market-based explanations of the supply of low-income housing rely on arguments about the "filtering" or "trickling down" of housing units from upper-income to lower-income groups. These arguments postulate that vacancies in the existing housing stock are created by moves of upper-income groups into newly constructed units, which are more desirable to these groups than existing units because of their physical condition and features or their relative price. These initial vacancies are filled by moves by members of the next lowest-income group, who take this opportunity to upgrade their own housing and thereby create subsequent vacancies, which are filled by the next poorest group, and so on. Housing units thus "filter down" from upper-income to lower-income groups as they age and presumably deteriorate in relative quality. To the extent that units decline in relative price (or absolute incomes rise) more rapidly than they deteriorate in quality, lower-income groups can upgrade the quality of their housing without increasing the fraction of their incomes spent on shelter (Quigley, 1979).

Within this context, housing policies of the type generally favored by the Reagan administration attempt either to increase the aggregate "flow" of units being filtered or to raise the level of the process at which lower-income groups can bid for units. Attempts to subsidize the purchase of new homes by either reducing building costs or increasing tax incentives for home ownership presumably have the effect of increasing the number of vacancies created at the top of the filtering chain and hence the number of units filtering down to lower-income groups, while housing allowances or other earmarked transfer payments serve to increase the quality of the units that subsidized households can afford at a given income. Arguing that there has been a long-term decline in the number of households occupying housing units with severe structural problems, such as the absence of inside plumbing, Reagan administration spokespeople have contended that the major remaining housing problem for the poor is affordability, or limited income, and that a system of housing allowances, together with some changes in the Community Development Block Grant Program to permit the limited construction of new housing is sufficient to meet the major housing needs of the poor (U.S. President's Commission on Housing, 1982, pp. 17-30).

As a corollary, the Reagan administration has actively pursued attempts to eliminate existing subsidy programs. Three of the last four administrations have been hostile to traditional housing subsidy programs and instead have supported increased reliance on the private market to supply housing for the poor. Beginning in the late 1960s, the emphasis on public housing gave way to various forms of subsidies provided to housing units and households in privately owned and operated buildings. Programs to provide housing subsidy certificates (Section 8 Existing) grew but there were still significant programs for new construction or substantial rehabilitation of low-income housing. Under the Reagan administration, the policy change has been pushed further. Subsidized construction for families has almost disappeared and the substantial rehabilitation portion of Section 8 was repealed in 1981. Rents were also raised substantially for low-income tenants in subsidized units. There has also been continuing pressure to lower the level of the rent subsidy to the families living in private housing. The basic philosophy was set out in the report of the President's Housing Commission in 1982:

> Nothing works unless the private sector works. . . . Since the 1960s, every government-stimulated burst of housing activity has been followed by a deep decline after the initial shock of the program has been absorbed, because the health of the private economy was being sapped by inflation. (p. 35)

In each of its budget proposals, the administration has advocated virtually no new construction or substantial rehabilitation for subsidized families and decreasing real rent subsidies. At the same time, it has proposed other policy changes undermining the economic situation of poor households. Even though Congress has rejected a number of proposed cuts, the aggregate effect of six Reagan budget cycles on expenditures for subsidized housing has been very large indeed.

President Carter's last budget provided for 255,000 additional units of subsidized housing in the 1981 fiscal year. The Reagan administration made two major cuts that year and has succeeded in virtually eliminating all construction after that. The fiscal year ending in October 1985 produced only about 20,000 new subsidized starts, primarily for the elderly and handicapped, but not for families. Further deep cuts were proposed for fiscal year 1987. Very few areas of the federal budget have absorbed cuts even approaching the magnitude of those in assisted housing programs (U.S. Dept. of Housing and Urban Development, 1981, 1986).

In addition to reducing the number of assisted households, the administration has proposed to reduce drastically the size of the subsidy to those households that continue to receive assistance. Draft regulations proposed in early 1986 by the Department of Housing and Urban Development (HUD) for the Section 8 Existing Housing program, the principal remaining subsidy program for families, called for substantial decreases in the maximum permitted rent levels for subsidized families using private housing.

The major thrust of recent federal housing policy, in short, has been to rely much more heavily on the free market "filtering" process, rather than government subsidy programs, to supply housing for poor and minority groups. The balance of this chapter evaluates the likely consequences of this shift on housing for these groups in the Chicago area. We look first at recent demographic and economic trends in the Chicago metropolitan area, then examine success of lower-income and minority groups in gaining access to adequate and affordable housing in this context.

THE CONTEXT FOR HOUSING POLICY: THE MOVEMENT OF PEOPLE AND JOBS

Like other large metropolitan areas, Chicago and its suburban ring have undergone vast changes since World War II, changes that shape many of the housing problems apparent today. The city reached its peak population, 3.6 million, in the 1950 census, a relatively small increase over the 3.4 million attained in the 1930 census. The central city's dramatic growth had occurred between 1870 and 1930. In 1950 the 1.56 million suburbanites were only 30% of the metropolitan total. As the city began to shrink in population, the suburban population soared to 2.7 million in 1960, 3.6 million in 1970, and 4.1 million in 1980, when 58% of the metropolitan population was suburban. Put another way, the suburbs grew 72% in the 1950s, 35% in the 1960s, and 14% in the 1970s (Chicago Fact Book Consortium, 1984, p. xvi). Between 1950 and 1980, the city lost 616,000 people, while thc suburbs gained 2,542,000 (Chicago Fact Book Consortium, 1984, p. xvi). During these three decades, the housing market had to create additional suburban housing for a population more than four-fifths the total in the central city, while those needing housing in the city dropped by almost one-sixth.

These aggregate population changes also reflect a large movement of the white population out of Chicago, and large increases in the black and Hispanic populations. Table 7.1 displays changes in overall population and for major racial and ethnic groups for the city between 1970 and 1975. This decline was the result of a major decline in the city's white population that was only partially offset by increases in the numbers of blacks and Hispanics. According to these figures, the city lost almost one-third of its white population between 1970 and 1980. Further, since many Hispanics are also classified by the Census Bureau as whites, these figures may understate the true level of white population decline.[1] This decrease in white population, which appears to have continued, albeit at a significantly lower rate, into the 1980s, has been accompanied by an increase in the city's black and particularly in the Hispanic populations. The city's black population, which had grown rapidly after World War II, largely as a result of migration from the South, grew at a much slower rate than the postwar period. This slowdown likely reflected this decline in in-migration as well as a slight increase in the movement out to a small number of majority black suburbs on the city's South and West sides.

The growth of the city's Hispanic communities has been far less visible but even more rapid. The Hispanic population reached 14% of

TABLE 7.1
Chicago Population Change by Race, 1970-1984

	1970	*1975*	*1980*	*1984*
White	2,223,775	NA	1,512,452	NA
Black	1,145,581	NA	1,197,220	NA
Hispanic[a]	249,332	NA	423,715	NA
Total	3,369,357	3,099,391	3,005,072	2,992,000

a. In the Census Hispanics can be counted as members of any race, so white Hispanics are counted under both categories. In addition, the Census Bureau changed procedures for counting Hispanics between 1970 and 1980, which may account for some of the growth in numbers.

the city's population and 4% of suburban residents by 1980 after a decade of very rapid expansion. Hispanics living in the city in 1980 had an average age of 22, compared to 25 for blacks and 39 for whites. They were three times as likely as white Chicago households and 63% more likely than blacks to have preschool-age children, suggesting that they would become a steadily growing fraction of the city's population, even without continuing immigration (Orfield & Tostado, 1983, pp. 18-23).

This decentralization of population was also paralleled by a substantial decentralization of employment. Table 7.2 displays changes in employment in the city and its suburbs' four major industrial groups between 1972 and 1982. The city's share of metropolitan area employment fell sharply in all four categories, and declined overall from 49% in 1972 to less than 40% in 1982. While these figures overstate the decline in Chicago employment because the 1982 data were collected in the midst of a major recession, more recent information suggests that the employment growth stemming from the recent recovery has been overwhelmingly concentrated in the suburbs. The first small increase in total employment in the city in the 1980s occurred in 1984, but this increase was less than a tenth of that in the suburban ring (Goff & Miller, 1986).

The decentralization of employment has had predictable effects on the economic disparity between the city and its suburbs. Table 7.3 displays overall median family income for the city and suburbs and for the city's major racial and ethnic groups between 1975 and 1983. Median city family income fell from 59% of the suburban level in 1975 to 52% eight years later. This gap widened even more sharply for blacks. In 1975, median black family income in Chicago was 68% of the suburban level, but fell to 50% in 1983, possibly as a result of the recession and reductions in transfer payment programs and an increase in middle-class black suburbanites. The gap between city and suburb grew,

TABLE 7.2
Employment in Four Major Sectors in Chicago and Suburbs, 1972-1982, in 1,000s

	Chicago	*Suburban Cook*	*Collar Counties*	*Percentage in Chicago*
Manufacturing				
1972	430.6	338.0	170.9	43.5
1977	336.0	351.5	196.9	38.0
1982	277.0	278.7	188.9	37.2
Retail				
1972	193.1	139.6	97.0	44.9
1977	178.3	178.8	118.7	37.5
1982	147.2	177.0	130.7	32.4
Wholesale				
1972	101.2	62.8	25.1	53.5
1977	91.2	79.2	35.9	44.2
1982	74.1	90.5	47.1	35.0
Service				
1972	157.3	51.2	36.6	64.2
1977	158.0	71.6	44.7	57.6
1982	215.2	125.0	84.1	50.6

SOURCE: Census of Business, 1972,1977, 1982.

however, among all racial groups over this period.

These changes in population and income—a decline in the number of households seeking housing within Chicago and a decline in relative income among those families still remaining—should have had considerable effects on the availability and price of housing inside the city. We turn next to an examination of these effects.

A naive application of the "filtering" process model for generating low-income housing to demographic conditions in Chicago over the last fifteen years would produce predictions of a substantial improvement in the quality and price of housing occupied by lower-income groups. Given average household sizes of 2.7-3 persons, the net population loss of 377,000 in Chicago between 1970 and 1985 should have produced between 125-140,000 new vacancies, an amount equivalent to more than 10% of the city's year round units. The creation of such a large volume of vacancies should have enabled a considerable number of lower-income and minority households to upgrade their housing at minimal cost.

The precise number of units made available for occupancy by lower-income groups would depend on the age of the newly vacated units, their initial quality, subsequent maintenance spending by landlords and owners, changes in the absolute and relative income of different racial

TABLE 7.3
Change in Median Family Income in Chicago and Suburbs, 1975-1983, by Race

	Chicago			*Suburbs*		
	1975	1979	1983	1975	1979	1983
White	12,000	16,100	19,679	17,500	24,100	29.252
Black	7,500	11,780	10,000	11,103	19,500	20,000
Hispanic	9.292	12.750	13,510	14,050	18,066	20,000
Total	10,122	14,000	14,650	17,150	24,000	28,330

SOURCE: Annual Housing Surveys.

and income groups, as well as a variety of other variables. Even with the most pessimistic assumptions about these variables, however, it is difficult to escape the conclusion that a properly functioning housing market would have produced a considerable improvement in the quality and cost of units occupied by lower-income and minority households, particularly in the last half of the 1970s. While one might expect a reduction in this improvement over the early 1980s because of a decline in out-migration, sharp declines in new construction in the suburbs stemming from the recession of 1981-1983, and a reduction in the absolute income of many minority households resulting from the recession and cuts in federal income support programs, on balance, Chicago's lower-income population should be better housed in the middle 1980s than it was fifteen years earlier.

There is indeed a variety of evidence suggesting that more housing became available over the 1970s in both Chicago and its suburbs. Table 7.4 reports changes in the number of available year round housing units and renter occupied units between 1970 and 1980 for Chicago and suburban areas. These figures show a large increase in new units in the suburbs and a decline in the city, but one that is smaller than the decline in population—and by inference, the number of households—that occurred over this same period. Even with the decline in the number of rental units in the city, the large out-migration of upper-income groups for new units in the suburbs should have expanded the supply of units available to lower-income groups.

There is also more direct evidence of this expansion. Table 7.5 reports on rental vacancy rates between 1975 and 1983 in both city and suburbs for all rental units, for those asking contract rents equal to or less than the average rent paid by poor Chicago families, and for those accessible to a moderate-income Chicago family willing to spend 30% of its income for rent. These data indicate that rental opportunities for lower-income

TABLE 7.4
Change in Year-Round Units and Renter-Occupied Units, 1979-1980, Chicago and Suburbs

	Percentage Change, 1970-1980	
	Year Round Units	*Renter Occupied Units*
Chicago	−2.7%	−9.8%
Suburbs, Total	*35.1%*	*41.8%*
Suburban Cook	26.8	30.2
DuPage	64.9	105.2
Kane	27.8	23.1
Lake	37.1	89.2
McHenry	48.9	29.3
Will	49.0	37.7

SOURCE: Bureau of Census, 1977, 1983.

TABLE 7.5
Rental Vacancy Rates, Poverty, and Low/Moderate-Income Levels, Chicago and Suburbs, 1975-1983

	1975		*1979*		*1983*	
Income Level	*Chicago*	*Suburbs*	*Chicago*	*Suburbs*	*Chicago*	*Suburbs*
Poverty[a]	7.4%	3.8%	7.1%	7.7%	8.5%	5.5%
Low/Moderate[b]	6.6	5.7	6.9	6.3	8.6	6.4
Overall	6.7%	7.3%	7.2%	7.3%	8.6%	6.7%

SOURCE: Annual Housing Surveys.
a. Vacancy rate among units asking contract rent at or below mean contract rent paid by families with income below poverty level in indicated year.
b. Vacancy rate among units asking contract rent at or below 30% of Chicago low and moderate family income in given year. Low and moderate family income is defined as 80% or less of median family income.

groups increased substantially in the city and moderately in the suburbs. Vacancy rates among units with contract rents accessible to both poor and moderate-income families increased to 8.5% in the city by 1983, and to 5.5% in the suburbs. Given these market conditions, it is reasonable to expect that housing conditions for lower-income groups should have improved over this period.

In fact, no such improvement has occurred. A close examination of the housing conditions and costs of lower-income and minority households in Chicago through the middle 1980s indicates that these households benefited only slightly from the loosening of the housing market through the middle and late 1970s, and that they were severely affected by the deterioration in both housing and employment markets in the early 1980s. By contrast with conventional expectations, poor households in general, and black and Hispanic households in particular, appear to be less well housed in the 1980s than before. The following sections of this chapter present evidence to support this contention. Particular emphasis is placed on levels of home ownership; the rental housing market; the sources, both market and nonmarket, of interracial differences in housing quality and cost; and the special problems of female-headed households. We then draw some conclusions about the likely efficacy of market-based housing policies in Chicago and similar cities.[2]

HOME OWNERSHIP FOR BLACKS, WHITES, AND HISPANICS

Home ownership in the United States, strongly supported by tax and other policies, increased from 44% of households in 1940 to 62% in 1960 to 65% in 1979. The percentage of black households owning their homes increased from 38% in 1960 to 44% in 1979 (*Statistical Abstract,* 1981, pp. 763-764). The Annual Housing Survey data for metropolitan Chicago shows that this trend continued for all three population groups in the late 1970s. The 1979-1983 period, however, brought a dramatic reversal of the trend, particularly for blacks. This trend has serious consequences both for the financial situation of families and for their access to education and jobs, since home ownership is the principal source of wealth for American families and since many neighborhoods with good schools and abundant job opportunities are open only to homeowners.

During the 1970s, minority households should have been able to increase their rate of home ownership appreciably. First, market conditions should have made Chicago city units relatively more attractive than suburban units than they might otherwise have been. As in most metropolitan areas, housing prices and mortgage interest rates increased sharply in Chicago over this period, particularly in the last half of the decade and particularly in the suburbs. Outside of Cook County, the median value of suburban owner-occupied housing more than tripled during the 1970s, and there were substantial increases in both down payment requirements and mortgage interest rates. These increases may have put most suburban units out of reach for many potential buyers, particularly first-time buyers with no accumulated equity in a previous home that could be drawn on as a potential down payment. In Chicago, by contrast, housing values did not appreciate as rapidly over this period. The Chicago housing inventory is considerably older and composed of smaller and less expensive units than the suburban housing stock. The large scale out-migration in the early 1970s may have exerted a downward pressure on prices that kept a large number of units within the reach of buyers who otherwise would have located in the suburbs. In short, it is reasonable to suppose that homes in Chicago were reasonable in price and in abundant supply compared to suburban units.

Minority households should have been particularly well placed to take advantage of this opportunity. Median family income among blacks and Hispanics grew at a relatively rapid rate over this period, especially in the last half of the decade. According to data from the Annual Housing Survey, median black family income in the city of Chicago grew from 62% of median white family income in 1975 to 73.2% in 1979, if cash transfer payments are included as income. Hispanic family income grew more slowly, but approached 80% of white median family income in 1979. While in both cases, these incomes were below those required to purchase the median-valued house in the city, they are much closer than in previous decades, suggesting that home ownership was financially possible for a larger share of minority households.

Available data indicate that while ownership did increase in Chicago over this period, there was only a slight increase among minority groups. According to U.S. census figures, the number of owner-occupied units inside the city increased by 7.2% between 1970 and 1980, or approximately 25,000 units. Table 7.6 displays the incidence of home ownership for each of the city's three major ethnic and racial groups between 1975 and 1983. These data indicate that the bulk of this increase was among white and Hispanic households. Home ownership increased among all three groups between 1975 and 1979, but the increase among blacks was only about half that among the other two groups. These figures also suggest that this increase was short-lived. Home ownership fell sharply among both blacks and whites between 1979 and 1983, while declining slightly among Hispanics. Among white households, home ownership reverted to levels slightly below 1975, while falling well below 1975 rates for blacks.

The causes for these group differences are difficult to determine precisely. Housing prices and mortgage interest rates increased very sharply over this period, and median family income in the city grew only

TABLE 7.6
Percentage of Chicago Households Owning or Buying Homes, by Race/Ethnicity 1975-1983

	1975	*1979*	*1983*
Race/Ethnicity			
White	47	53	46
Black	40	43	34
Hispanic	24	31	31

SOURCE: Calculations from Annual Housing Survey files.

slightly, if at all. Many households that otherwise would have purchased homes over this period were unable to do so. Further, given the high unemployment rate in Chicago stemming from the recession of 1981-1983, lower-income homeowners may have experienced extended periods without income during this period that compelled them to default on their mortgages. This factor may have been particularly significant for blacks, for whom unemployment rates were at or above 20% for much of this period.

The extent to which these trends have persisted through the recovery in the mid-1980s is difficult to gauge. Mortgage interest rates have only recently begun to decline appreciably, and minority unemployment and income have remained depressed. One might hence expect some increase in home ownership levels among white households, who have likely been the most directly constrained by interest rates, but little if any improvement among blacks.

HOUSING QUALITY AND OVERCROWDING

A general improvement in housing quality among lower-income and minority groups in Chicago over this period might also be expected. An increase in the number of units entering the filtering process should allow lower-income groups to afford a better-quality unit for essentially the same share of their income, as was the case over the late 1970s. Absolute incomes were rising in the late 1970s, therefore likely increasing the rate at which structurally deficient units were removed from the housing stock by demolition. Similarly, the increased availability of housing should permit households that have been "doubling up" to seek their own units, thereby leading to reduced overcrowding.

There is, in fact, little evidence of any appreciable improvement in either the quality or the congestion of housing units occupied by low-income and minority groups. Table 7.7 presents data on the fraction of households of different income and racial groups occupying housing either in need of rehabilitation or overcrowded. Need for rehabilitation is defined by a measure developed by the Congressional Budget Office for units that lack complete plumbing or kitchen facilities or have multiple minor maintenance problems. Units are classified as overcrowded if they are occupied by more than one person per room, excluding kitchen and bathroom, or by more than two persons for each bedroom.

These figures suggest that there has been little change, either relatively or absolutely, in the quality of housing occupied by minority groups over this period. At any given income level, black and Hispanic households occupy "problem" units at rates between one and a half and seven times those of whites. Interracial disparities in overcrowding narrow slightly in 1979, but increase sharply by 1983.

Perhaps more important, there appears to be little difference between the quality of the housing occupied by poor white households and more prosperous minority families, who presumably can afford higher quality units. There is little difference between the incidence of unit in need of rehabilitation among black and Hispanic households with incomes above the poverty level and white households with incomes below it; and better-off minority households are substantially more likely to occupy overcrowded dwelling units than poor white ones.

RENT BURDENS: PATTERNS BY RACE

There is also little evidence of any lessening of rent burdens among lower-income and minority groups in Chicago. Table 7.8 displays contract and gross rent to income rations for residents of unsubsidized housing by poverty status and race/ethnicity over this period. These figures indicate that the share of income that poor households in general, and black households in particular, have spent for housing increased sharply between 1975 and 1983. Rent/income ratios for households with income below the poverty level stayed roughly the same between 1975 and 1979, then increased sharply between 1979 and 1983. This increase was particularly large for black households, who experienced an increase of 11% in the fraction of their income they paid to landlords, and an increase of 15% in the share of income devoted to gross rent, or total housing expenses. On average, poor black households

TABLE 7.7
Percentage of Households Occupying Housing Units in Need of Rehabilitation or Overcrowded, by Race/Ethnicity and Poverty Status, 1975-1983

	1975		*1979*		*1983*	
Race/ Ethnicity	*Below Poverty*	*Above Poverty*	*Below Poverty*	*Above Poverty*	*Below Poverty*	*Above Poverty*
Units in Need of Rehabilitation[a]						
White	11	4	9	5	10	5
Black	25	12	25	9	22	11
Hispanic	12	10	18	12	18	9
Units Overcrowded – 1+ Persons/Room						
White	5	3	5	2	6	1
Black	18	8	14	7	15	6
Hispanic	22	14	18	17	29	12
Units Overcrowded – 2+ Persons/Bedroom						
White	5	4	5	3	5	3
Black	17	10	16	9	21	9
Hispanic	25	18	23	19	35	17

SOURCE: Calculations from Annual Housing Survey files.
a. A unit is classified as in need of rehabilitation if it has at least one major deficiency or if it has two or more secondary defects. The major deficiencies are (1) the absence of complete plumbing facilities and (2) the absence of complete kitchen facilities. The secondary defects are (1) three or more breakdowns of six or more hours each time in the heating system during the last winter; (2) three or more times completely without water for six or more hours each time during the prior 90 days, with the problem inside the unit; (3) three or more times completely without flush toilet for six or more hours each time during the prior 90 days, with the problem inside the unit; (4) leaking roof; (5) holes in interior floors; (6) open cracks or holes in interior walls or ceilings; (7) broken plaster or peeling paint over greater than one square foot of interior walls or ceilings; (8) unconcealed wiring; (9) the absence of any working light in public hallways for multiunit structures; (10) loose or no handrails in public hallways for multiunit structures; and (11) loose, broken, or missing steps in public hallways in multiunit structures. Units lacking complete plumbing facilities or with four or more other defects or deficiencies are classified as in need of substantial rehabilitation.

were paying almost 80% of their income for housing in 1983.

While much of this increase in relative rent burdens among poor and black households is the result of declines in income stemming from the recession of 1981-1983, and reductions in transfer payments, there is also evidence that absolute rents increased at a more rapid rate over the early 1980s than earlier. Overall, rents in Chicago increased at a compounded annual rate of 9% between 1979 and 1983, a rate of increase over 50% higher than the previous four years, when rents grew at a compounded annual rate of 5.9%. The most rapid growth in 1979-1983 occurred at the top end of the market, where rents grew at a

TABLE 7.8
Mean Contract Rent/Income and Gross Rent/Income Ratios,
Unsubsidized Units by Race/Ethnicity and Poverty Status, 1975-1983

	Below Poverty			*Change*	*Above Poverty*			*Change*
Race/Ethnicity	*1975*	*1979*	*1983*	*1975-1983*	*1975*	*1979*	*1983*	*1975-1983*
Contract Rent/Income[a]								
White	62%	61%	66%	+4%	20%	21%	24%	+4%
Black	56	57	67	+11%	20	19	24	+4%
Hispanic	53	50	57	+4%	16	16	18	+2%
Gross Rent/Income[b]								
White	67%	68%	74%	+7%	22%	24%	27%	+5%
Black	64	67	79	+15%	23	23	28	+5%
Hispanic	61	61	69	+8%	18	20	22	+4%

SOURCE: Calculations from Annual Housing Survey file.
a. *Contract Rent.* Contract rent is the monthly rent agreed to, or contracted for, even if the furnishing, utilities, or services are included.
b. *Gross Rent.* Gross rent is the contract rent plus the *estimated average monthly cost of utilities* (electricity, gas, water) and fuels (oil, coal, kerosenes, wood, etc.) if these items are paid for by the renter (or paid by someone else, such as a relative, welfare agency, or friend) in addition to rent.

compounded annual rate of 10%; but the average rent for less expensive units also grew at more rapid rates then during 1975-1979. Among lower-income and moderate-income groups, rents paid by whites grew at a higher rate than those paid by blacks during this earlier period; but black rents increased more rapidly than whites during the early 1980s.

THE IMPACT OF THE HOUSING CRISIS ON FEMALE-HEADED HOUSEHOLDS

In describing a housing crisis affecting the poor, one is also describing a housing crisis that very disproportionately affects women and their children. Female-headed households accounted for 56% of the black households in Chicago in 1983, and three-fifths of them lived below the poverty line. There were also large numbers but a much smaller percentage of white female-headed households below poverty (see Table 7.9).

Households headed by women living below the poverty line faced extraordinary rental burdens, whether they were white or black. On average, their gross rent (including utilities) in 1983 was more than three-fourths of their cash income. The situation was slightly less desperate than these figures suggest because of noncash income in the form of food stamps, Medicaid, and so on, for some of these families. Nevertheless, it is apparent that many had virtually no cash for clothing, transportation, school materials, and other necessities once the rent was paid. These figures certainly help account for both the frequency of nonpayment and eviction problems in very poor neighborhoods and the inability of poor families to meet the cost of other basic necessities. Even if one assumes that there was a significant amount of unreported income, these households faced extraordinary financial problems.

In contrast, only 3% of white male-headed suburban households had incomes below the poverty level in each year. On average, white male suburban households above poverty paid 20% of their income on gross rent in 1975 and 23% in 1983.

CONCLUSIONS: RACE AND HOUSING IN CHICAGO

The results of this analysis are striking. In spite of housing market conditions that should have favored a substantial expansion of home

TABLE 7.9
Female-Headed Households and Housing Problems, 1975 and 1983 (in percentages)

	1975		*1983*	
	City	*Suburbs*	*City*	*Suburbs*
Female households				
Black	45	25	56	33
White	30	17	39	24
Female households below poverty				
Black	52	38	59	40
White	23	14	19	14
Income spent on gross rent, unsubsidized families below poverty				
Black	64	67	79	82
White	69	74	75	86

SOURCE: Annual Housing Surveys.

ownership among minority households during the 1970s, there is little, if any, evidence that low-income and minority households are any better housed in the middle 1980s than fifteen years earlier. Levels of home ownership among blacks have declined over this period, and the incidence of substandard housing and overcrowding among lower-income black and Hispanic households has either remained static or increased. In addition, these households are paying more, both in absolute terms and relative to income, for housing compared to other groups than earlier. These groups appear to have benefited slightly, in terms of reduced growth in rent payments and some increases in the rate of home ownership for Hispanics, from the loosening of the housing market during the middle 1970s. But these gains appear to have been small and temporary. In sum, it appears that lower-income and minority households are paying more for housing that is no better and, in many cases, appreciably worse than the units that they occupied fifteen years ago.

THE CAUSES OF MARKET FAILURE: HOUSING MARKET RACIAL SEGMENTATION

Discussions of housing in metropolitan Chicago often presume the existence of a single large metropolitan housing market. In a certain sense, of course, this is true. Housing across the entire metropolitan area, for example, is not typically sold through markets that terminate at municipal boundary lines. The same powerful commercial media, particularly the dominant metropolitan dailies and the electronic media, are used to market housing across the entire region. It is true that investors and institutions often operate simultaneously in widely dispersed sectors of the metropolitan market. These actors include lending institutions, real estate firms, relocation services, developers, and many others. Millions of people are willing to look for their next housing unit some distance from their present home, and it is commonplace to cross not just one but several jurisdictional boundaries every day in traveling from home to work. Both consumers and producers operate on a metropolitan basis. There are, in other words, some basic characteristics of a single market.

There is, however, one fundamental error in this description. Within the large housing market area there is extreme separation between whites and blacks and between Hispanics and blacks, only a small portion of which can be explained by income differences. Hispanics and whites are also highly segregated from one another. Incomes among these groups are strikingly different on average, of course, but both the overlap in the income distribution and the range of housing costs within various sectors of the market would produce multiracial communities in virtually all areas if income were the determining factor. In fact, there is powerful evidence that the basic causes are present discrimination and the continuing effects of past residential isolation.

Chicago has been near or at the top of the list of segregated housing markets for decades, and the 1980 Census showed no significant change in the overall pattern (Taeuber & Taeuber, 1965). Karl Taeuber's calculations of segregation levels in 1980 in all major U.S. cities found Chicago the most segregated. Using his dissimilarity index, on which total definition of areas on racial lines with complete separation between two groups is 100 and random distribution of both groups is 0, the metropolitan Chicago score in 1980 was 92; the level was slightly higher in the city. The typical black family in Chicago lived on a block that was 96% minority (Hogan, 1983; Taeuber, 1983).

Five of every six blacks in the metropolitan area depended on the city's intensely segregated housing market in 1980, and the rapidly growing number of black suburbanites were concentrated in a small number of suburbs, including some that were being incorporated into ghettos expanding from the city's South and West sides. By 1980 almost two-fifths of black suburbanites lived in majority black suburbs and many others in suburbs undergoing racial change.

Hispanics were almost as segregated from blacks as were whites, and the Hispanic level of segregation from non-Hispanic whites was higher than that experienced by earlier immigrant groups. The dissimilarity index for white- Hispanic segregation in the metropolitan housing market in 1980 was 70; some subgroups were even more segregated (Hogan, 1983; Orfield & Tostado, 1983).

Predominantly minority areas and areas undergoing rapid racial transition were much worse off in terms of economic and educational measures than predominantly white or stably integrated areas within the housing market. There were, in other words, clear costs associated with the processes of neighborhood segregation (Orfield, Woolbright, & Kim, 1984).

The segregation levels were not the result of voluntary preference of minority households for racial separation but were the result of present and past discrimination. Surveys of blacks and Hispanics in the metropolitan area show that both groups perceive strong continued discrimination in the housing market and that those who earn more actually perceive more not less discrimination (*Chicago Sun Times*, October 16, 1983). This impression was confirmed by economist John Kain's analysis of the relationship between income and location by race within the metropolitan housing market. The analysis of Chicago market data showed that 73% of blacks lived in five Annual Housing Survey districts that were more than 65% black. Only 25% would have been expected to live in these areas on the basis of the economic match between their income and the local housing market. On purely economic grounds there should have been about three times as many black suburbanites as were actually reported (Kain, 1984, p. 239). Tests by fair housing groups in several sectors of the Chicago housing market in the mid-1980s disclosed many examples of overt discrimination and steering (Hall, Peterman, & Dwyer, 1983; Leadership Council for Metropolitan Open Communities, 1985).

The Chicago housing market, in other words, is not a true single market. With relatively few exceptions, there is a separate housing market for blacks. Most Hispanics also live within separate areas. Minority households have negative expectations about fairness in the housing market and there is evidence that discrimination continues to exist to a serious degree. It is clear that equal income does not produce a similar range of housing choices across racial lines and that the income differences between blacks and whites do not explain the extreme separation. In fact, rather than income differences explaining segregation, the process may work somewhat in the opposite direction. Poor people have geographic limitations to education, transportation, and employment.

POLICY IMPLICATIONS

These results suggest that the Chicago housing market has not functioned efficiently in the allocation of housing units to minorities and lower-income groups. Units are available to these groups within constraints other than income and style preferences. Because of the extreme segregation in the city's housing allocation system, blacks and, to a lesser extent, Hispanics have had their housing choices severely constrained and been compelled to pay premium rents for relatively low-quality housing.

This finding casts strong doubt on the likely efficacy of proposals to rely on the "private" market to improve the housing of low-income and minority groups in Chicago and other cities. Attempts to increase the volume of housing "filtered down" to lower-income households by stimulating overall levels of housing construction are likely to be ineffective in either improving the quality or cost of lower-income housing, particularly that available to blacks. The extreme segregation of the city's housing market has prevented units from "filtering" down to lower-income groups by making it effectively impossible for minorities, particularly blacks, to rent or buy units that they can demonstrably afford. Blacks have been confined to an artificially narrow range of housing choices—in effect, a second housing market—and have been required to pay increasing rents in this market despite of increasing vacancies in the larger, white housing market. For more prosperous blacks, the major effects of this restriction appear to have been an underconsumption of housing in the form of lower levels of home ownership or occupancy of poorer-quality rental units than similarly situated white households. Poor black households have experienced larger rent increases relative to income than other poor households.

Hispanic households appear to have been moving into units available in previously all-white neighborhoods at relatively low prices. Hispanics appear to have had somewhat greater success than blacks in buying homes and have consistently paid lower rents relative to income than either blacks or whites. These results are consistent with the notion that landlords in previously all-white areas have been willing either to leave available units vacant or sell or rent to Hispanics, even at a loss, rather than make these units available to blacks.

These findings suggest, in brief, at least in Chicago and other cities with similar housing markets, that the major housing "problem" for lower-income and minority households is not only an adequate supply of housing available at affordable rents, but also securing access to those units. Without aggressive enforcement of open housing or other policies intended to ensure that all groups have access to those housing units that they can afford, increases in the overall rate of residential construction are unlikely to have much effect on how well poor and minority households are housed and how much they pay for shelter.

These results also indicate the limited effectiveness of housing allowances or other demand subsidies by themselves in expanding the housing choices of low-income households without related measures to ensure that these households in fact have access to all available units. While large numbers of poor Chicago households of all three major racial or ethnic groups are paying excessive shares of their incomes for housing, much of this burden stems from restrictions on the number of units that these households can rent. The extreme segregation of the city's housing market serves to tighten artificially the portion of the market available to blacks and increase rents far above what they would otherwise be if access were broader. Mere expansion of rent subsidies or other demand subsidies would not solve the problem.

These results also indicate that Reagan administration policies of reducing housing subsidies and the lack of enforcement of fair housing laws are likely to worsen the housing conditions of minorities and lower-income families. Without some considerable financial support, these households are unlikely to be able to afford most suburban units. Without strong enforcement of open housing requirements, they are unlikely to be able to rent such units even if they could afford them.

The importance of pursuing these goals together is demonstrated by local evaluations of the likely consequences of HUD's proposals to reduce the maximum rent payable under the Section 8 Existing Housing Program. A 1986 study of rent levels of two-bedroom units in apartment complexes containing almost 30,000 apartments reported that HUD's existing Fair Market Rent for such units was already almost 11% below the actual market price within the Chicago metropolitan area. Moreover, major Chicago landlords were in the process of implementing substantial rent increases that would widen the gap. HUD proposed, however, to lower the Fair Market Rent substantially, further limiting the possibility of finding a unit, particularly in the areas where the new jobs were being created.

The Leadership Council for Metropolitan Open Communities is a metropolitan Chicago fair housing organization. It was under contract with HUD to meet HUD's obligation under a court order to find housing outside ghettos for 7,100 central city poor families. It reported that the HUD proposals would have virtually destroyed its placement program had it been in effect in 1985. Instcad of placing 212 families in such areas, it could have placed only eight. In the two areas where the largest increases in jobs were occurring, DuPage and Northwest Cook counties, total placements would have dropped from 114 to 2. A separate study in DuPage County found that a survey of rents in apartment complexes with almost 16,000 units found that less than 1% of the one-bedroom and two-bedroom units would be affordable for subsidized families under the proposed regulations and 1% of the three-bedroom units (Hope Fair Housing Center, 1986). After successfully terminating programs to build or rehabilitate housing for low-income families, HUD was effectively constraining the private market rental program in a way that would reinforce segregation and virtually eliminate access by low-income people to the areas of the greatest economic, educational, and housing opportunities.

An alternative experiment program that combines housing subsidies with measures to expand the housing choices of low-income and minority households is already in operation in the area. Other such programs must be implemented. The failure to take innovative and remedial steps will leave the Chicago market highly segregated, and the burden of that market falling on the poor and minorities.

NOTES

1. This overlap stems from the Census Bureau's practice of classifying blacks and whites but not Hispanics as members of different races. By contrast, Hispanics are classified as members of distinctive ethnic groups who may be of any race. Depending on their country of origin, therefore, Hispanics may also be classified as either black or white.

2. The major data source for this analysis is the Annual Housing Survey, which is taken by the Census Bureau for the U.S. Department of Housing and Urban Development in a number of large cities at 3-4 year intervals. Surveys were performed in Chicago in 1975, 1979, and 1983; with over 15,000 housing units in the Chicago metropolitan area being surveyed in 1975 and 1979 and approximately 7,500 units in 1983.

REFERENCES

Busk, C. (1986, January 26). Rents going up, 5 to 10 percent hike forecast for area apartment dwellers in spring. *Chicago Sun-Times.*

Chicago Fact Book Consortium. (1984). *Local community fact book: Chicago metropolitan area 1980.* Chicago: Chicago Review Press.

Cutler, I. (1976). *Chicago: Metropolis of the mid-continent.* Dubuque, IA: Kendall Hall.

de Vise, P. (1976). Social change. In D. Simpson (ed.), *Chicago's future* (pp. 14-132). Champaign: Stipes.

Downs, A. (1983). *Rental housing in the 1980s.* Washington, DC: Brookings Institution.

Goff, L., & Miller, M. (1986, April 28). DuPage the arrogant. *Crain's Chicago Business.*

Hendrix, C. (1986, March 17). *Filing with HUD rules docket clerk on docket No. N-85-1573; FR-2133.*

Hogan, W. C., III, (1983). *Residential segregation in metropolitan Chicago: 1980.* Unpublished B.A. paper, University of Chicago.

Orfield, G., & Tostado, R. (Eds.). (1983). *Latinos in metropolitan Chicago.* Chicago: Latino Institute.

Quigley, J. (1979). What have we learned about urban housing markets? In P. Mieszkowski & M. Straszheim (eds.), *Current issues in urban economics* (pp. 391-428). Baltimore: Johns Hopkins.

Rosenbaum, J. E., Kulieke, M. J., Leon, Rubinowitz, & Louis, D. A. (1982). *The effects of residential integration on poor children.* Paper presented at the annual meeting of American Educational Research Association.

Testa, M., & Lawlor, E. (1985). *The state of the child: 1985.* Chicago: Chaplin Hall Center.

U.S. Bureau of the Census and Department of Housing and Urban Development. (1985). *Annual housing survey: 1983. Housing characteristics for selected metropolitan areas, Chicago Illinois* (Current Housing Reports H170-83-22). Washington, DC: Government Printing Office.

U.S. Department of Housing and Urban Development. (1981, March). *FY 1982 budget revised.* Washington, DC: Government Printing Office.

U.S. Department of Housing and Urban Development. (1986, February). *FY 1987 budget.* Washington, DC: Government Printing Office.

U.S. President's Commission on Housing. (1982). *The report of the president's commission on housing.* Washington, DC: Government Printing Office.

Wintermute, W. (1983). *Recession and recovery: Impact on black and white workers.* Chicago: Urban League Report.

8

Housing Discrimination in Small Cities and Nonmetropolitan Areas

JULIA L. HANSEN
FRANKLIN J. JAMES

VERY LITTLE RESEARCH has examined housing discrimination encountered by racial and ethnic minorities in nonmetropolitan areas and small cities. Nor has the impact of such discrimination on minority housing conditions been assessed. Given that minority populations are concentrated in metropolitan areas, the focus of previous research on larger cities is understandable. In 1980, for example, 82% of blacks, 88% of Hispanics, and 48% of Native Americans lived in standard metropolitan statistical areas. Nevertheless, very large minority populations reside outside these areas.

The Housing Market Practices Survey of the Department of Housing and Urban Development represents the most ambitious effort undertaken to assess geographic patterns of housing discrimination against blacks. This study performed audits of discrimination in 40 metropolitan areas, and concluded that "blacks encountered more discriminatory treatment, on average, in large SMSAs than they did, on average, in small SMSAs" (U.S. Department of Housing and Urban Development [HUD], 1979A, p. ES-14). This was found to be true in both rental and "for sale" housing markets. Unfortunately, these results have little real implications even for small cities. In the study, small metropolitan areas were represented by eight areas having central city populations of from 50,000 to 100,000 in 1970 (U.S. Department of Housing and Urban Development [HUD], 1979A). Overall metropolitan area populations were, of course, substantially larger.

The most significant studies of housing discrimination against Hispanics have also focused on large urban places (HUD, 1979b; Feins & Holshouser, 1984; James, McCummings, & Tynan, 1984). Audits similar to those used in the Housing Market Practices Study have been performed in Dallas, Denver, and Boston. Other kinds of analyses designed to measure evidence of housing discrimination against Hispanics have been done in Denver, Houston, and Phoenix. To our knowledge, no significant studies have been made of housing discrimination against Native Americans, a group much more highly concentrated in nonmetropolitan areas than is either of the other two groups.

This chapter provides preliminary documentation of patterns of housing discrimination against blacks, Hispanics, and Native Americans in small cities and nonmetropolitan communities in Colorado. Two types of evidence are presented. First, new audits of discrimination against these groups are reported for two of Colorado's small cities: Grand Junction and Greeley. Approximately 90 audits were performed in the rental housing markets of each city during the summer and fall of 1983. As will be seen, these audits provide both statistical and anecdotal evidence of the patterns and severity of housing discrimination. It should be noted that the audits of discrimination against Native Americans are unique. No other such studies have been done. Discrimination against each of the three minority groups is measured using the same research instruments and methods, thus permitting a comparison of the severity of the problems they encounter.[1]

Second, the extent of discrimination in nonmetropolitan communities is assessed by using econometric methods to measure minority-Anglo disparities in home ownership and in housing space. This econometric research is performed for nonmetropolitan communities in Colorado. For purposes of comparison, a comparable analysis is presented for the Denver metropolitan area. It is well established that such disparities provide useful indexes of the impact of discrimination on minority housing conditions in urban areas (Yinger, 1979). The chapter concludes with observations regarding the implications of the findings for fair housing policy in small cities and nonmetropolitan areas.

INTRODUCTION TO GRAND JUNCTION AND GREELEY

Both Greeley and Grand Junction are rapidly growing, urban communities. Grand Junction is the major urban center on Colorado's

Western Slope, and has functioned as an administrative center of Colorado's once burgeoning energy industry. Greeley is a community north of Denver and Boulder, on Colorado's Front Range, and is the site of the University of Northern Colorado. Other sectors of its economic base include the processing of agricultural products and some manufacturing activity.

The comparatively rapid growth of both communities is apparent in Table 8.1. Grand Junction's population grew from 20,000 in 1970 to 28,000 in 1980, or by about 40%. Greeley's population grew from 39,000 to 53,000 during the 1970s, or by 36%. By contrast, the population of Colorado grew by only 31% between 1970 and 1980. Audits were facilitated by the rapid growth of these communities because both communities were accustomed to dealing with newcomers.

The populations of both communities are predominantly white.[2] Fewer that 700 blacks live in the two cities; the total Native American population is slightly under 350 (Table 8.2). The bulk of the minority populations of Greeley and Grand Junction are persons of Spanish origin. Hispanics make up 9.5% of Grand Junction's population, and almost 16% of the population of Greeley.

Table 8.3 presents three basic indicators of the economic status of racial and ethnic groups. The evidence suggests that blacks suffer from the lowest socioeconomic status. In Greeley, the median household income of blacks was less than one-half that of whites in 1980, and their poverty rate was almost five times higher. Both Hispanics and Native Americans enjoyed higher household incomes and lower poverty rates compared to blacks in Greeley. Indeed, median household income of Native Americans is reported to have been over 40% higher among Greeley's Native Americans than among whites in that city. Much the same ranking appears to hold in Grand Junction. Poverty rates for individuals in 1980 were highest among blacks, and median *family* incomes were lowest.[3] However, Native Americans enjoyed only somewhat higher economic status than did Hispanics in Grand Junction.

Reflecting the small minority populations of the two cities, neighborhood segregation is relatively limited, though it clearly exists. In Greeley in 1980, for example, 18% of Hispanics occupied two census tracts where Hispanics make up more than half of the tract populations.[4] Two-thirds of this city's Hispanics lived in six tracts where Hispanics make up 19% or more of tract populations.[5] These same tracts contained only 28% of the city's overall population. Similarly, over half of Greeley's tiny black population lived in four tracts housing only 21% of overall populations.[6]

TABLE 8.1
Populations of Grand Junction and Greeley: 1970 and 1980

	1970	*1980*	*Percentage Change*
Grand Junction City	20,170	28,144	39.5
Greeley City	38,902	53,006	36.3
Colorado	2,209,596	2,889,964	30.8

SOURCE: U.S. Bureau of the Census.

Similar patterns of neighborhood segregation are apparent for Hispanics in Grand Junction. In 1980, 19% of Grand Junction's Hispanics lived in a single census tract where they made up almost two-thirds (65%) of the tract population.[7] Altogether, 26% lived in four tracts where they made up at least one-fifth of tract populations. No Grand Junction tract housed more than a handful of blacks or Native Americans, however.

There are additional reasons to believe that minorities face significant discrimination in these two cities. In 1981, the Colorado Civil Rights Commission held public hearings in Grand Junction. The hearings were intended to collect testimony on the degree of discrimination. Both civil rights organizations and individuals reported that employment and housing discrimination was frequent. However, hard evidence was generally lacking (Colorado Civil Rights Division, 1981). In 1984, Greeley newspapers reported situations on the campus of the University of Northern Colorado in Greeley that point to problems between some blacks and Anglo students. An in-depth assessment of the situation at the University reported that racial and ethnic antagonism was widespread between the city residents and minority university students. Some incidents of racial violence and cross burning have occurred. Moreover, the University itself is reported to be guilty of discrimination against women and minorities in hiring and promotion policies (McCummings, 1985).

Rental housing markets were "soft" with excess supply in both Greeley and Grand Junction during the time periods of the studies—summer and fall, 1983. Grand Junction's economy was severely depressed by shutdowns in the oil shale industry, as well as general economic recession. Greeley also experienced major economic setbacks. Market softness had two effects on the study:

TABLE 8.2
Racial and Ethnic Compositions of the Populations of Grand Junction and Greeley: 1980

	Grand Junction		*Greeley*	
	Number	*Percentage*	*Number*	*Percentage*
White	26,735	95.0	47,678	89.9
Black	182	0.6	486	0.9
Native American	189	0.7	158	0.3
Spanish Origin[a]	2,674	9.5	8,260	15.6
Total	28,144	100.0	53,006	100.0

SOURCE: U.S. Bureau of the Census.
a. Persons of Spanish origin may be of any race.

TABLE 8.3
Indicators of the Socio-economic Status of Racial and Ethnic Groups in Grand Junction and Greeley: 1980

	Median Household Income	*Percentage of Families in Poverty*	*Homeownership Rate (percentage)*
		Grand Junction	
White	13,280	8.5	57
Black	N/A	N/A	49
Native American	12,980	N/A	17
Spanish Origin[a]	11,140	23.3	45
		Greeley	
White	14,901	8.4	56
Black	7,040	41.4	16
Native American	21,250	24.1	36
Spanish Origin[a]	12,080	31.4	42

SOURCE: U.S. Bureau of the Census.
a. Persons of Spanish origin may be of any race.

—It assured an ample supply of advertised units for audits;
—it may have reduced agent or landlord interest in discrimination.

Discrimination against minorities is easiest when markets are "tight," demand exceeding supply, so that potential tenants are plentiful. Landlords in Grand Junction and Greeley had reason to fear protracted vacancy if they failed to rent to minorities.

Table 8.4 shows that 92 audits were conducted in the Grand Junction

TABLE 8.4
Number of Rental Audits in Grand Junction and Greeley

	Number of Audits	
Type of test	*Grand Junction*	*Greeley*
Black/Anglo test	32	25
Hispanic/Anglo test	31	36
Native American/Anglo test	29	34
Total audits	92	95

SOURCE: Center for Public-Private Sector Cooperation Audit, 1983.

area, and 95 in Greeley. For analytic purposes, an *audit* is defined as a matched pair of reports. Thus the 187 audits involved 374 separate visits by auditors with rental agents or landlords. Audits in which one or both auditors either failed in efforts to interview agents or landlords or failed to report their interviews accurately are not included in these counts.[8]

DISCRIMINATION IN GRAND JUNCTION

In Grand Junction, 93.5% of the agents or landlords that the auditors encountered were Anglo, 2.7% were Hispanic, 1.1% were Native American, and 2.7% were other minorities. There were no black rental agents. The virtual white monopoly on real estate sales positions increases the vulnerability of minorities to discrimination. Indeed, discrimination against blacks, Hispanics, and Native Americans was found to be common in Grand Junction.

Three primary types or forms of discrimination are measured in the audits:

(1) Availability—differences in the quality or quantity of information about available units provided to minority and Anglo apartment seekers by real estate agents and/or apartment managers.
(2) Terms and Conditions—differences in terms presented to minority and Anglo apartment seekers as a condition of renting a unit, that is, rent, security deposit, lease requirement, and so on.
(3) Salesmanship/Steering—practices employed by agents/apartment managers that encourage minorities or Anglos to choose a unit in certain minority, Anglo, or integrated areas.

INFORMATION ON HOUSING AVAILABILITY

Studies have shown that agents frequently conceal available housing from minorities (HUD, 1979a). Such concealment is difficult for minorities to detect, and is highly effective in turning minority applicants elsewhere. To determine the quantity of information provided to minority and Anglo auditors, auditors were directed

—first, to inquire about the availability of the *advertised* unit;
—second, to ask about units comparable to the advertised unit in cost, size, and location; and
—third, if fewer than three prospective units were identified, to ask about different types of units that might meet the auditors stated needs.

The advertised unit was available in Grand Junction for immediate inspection for virtually all auditors. Even so, instances of discrimination in the provision of this basic level of information on housing availability did occur. The following case is illustrative.

Case I

Mrs. X, a black female auditor, appeared at 9:00 a.m. to inquire about a single-family detached house that was advertised for rent. She was told that the apartment had already been rented, that there were no comparable apartments available and that there was no waiting list.

At 2:00 p.m. on the same day, the Anglo auditor appeared and requested information about the same house. Not only was the house said to be available, but the auditor was told about other houses that the agent knew about that were comparable to the advertised house. The agent also stated that he had to be selective about who he rented the house to. He stated that he had had a bad experience previously with a tenant who left owing several months rent.

The black auditor involved in this case was unaware of discrimination until she compared notes with the Anglo auditor, suggesting the substantial difficulty of detecting discrimination in the provision of information. Instances of such blatant discrimination were also encountered by Hispanic auditors. The following case is illustrative.

Case II

Mrs. X, a Hispanic female auditor, responded to the ad in the newspaper by calling the agent to make an appointment to get additional information. On the telephone, the agent was very enthusiastic about renting the unit, stating that it was very well kept and in a nice neighborhood. When the auditor appeared, the agent seemed somewhat surprised. The agent stated that she wasn't sure when the unit would be available—"maybe in a few days." The agent suggested that the auditor look at another in another neighborhood. Upon inspection, the auditor found the alternative unit to be dilapidated, dirty, and in a lower-income neighborhood. Mrs. Y, an Anglo female auditor, responded to the same ad a day later. Not only was the unit offered for rent, but the agent said that the rent and the deposit could be negotiated if the auditor would rent immediately.

Auditors were also instructed to inquire about other units that were comparable in terms of cost, size, and location to the advertised unit. Minority auditors were told of fewer comparable units than were Anglo auditors. For example, 72% of black auditors were told that no comparable units were available; only 56% of Anglo auditors were told of no such unit in these tests. Similarly, 62% of Native American auditors were told of no comparable units; the comparable figure for Anglo auditors was 52%.

Overall, Anglo auditors were told of more housing units that were "serious possibilities" for meeting their housing needs (Table 8.5).[9] In Grand Junction, this gap in basic information was largest for Native Americans. Native American auditors were told of 29% fewer possibilities than were their Anglo counterparts. In many cases, the Anglo auditors were asked for their telephone number in the event that something else became available; this was not the case with Native American auditors. Black auditors also ran into marked discrimination of this type. They were told of 17% fewer possibilities than were the Anglos involved in these same tests. However, the average number of "serious possibilities" identified for Hispanic auditors was essentially the same as that identified for Anglo auditors.

The overall similarity in the quantity of information provided to Hispanic and Anglo auditors may conceal differences in the kinds of information provided. For example, comparable audits of discrimination against Hispanics in the Denver metropolitan area in 1982 also found that much the same numbers of units were volunteered as serious possibilities to Hispanic and Anglo auditors—0.92 to Hispanics and 1.06 to Anglos, a statistically insignificant difference (James et al., 1984, p. 112). The Denver audit was designed to measure geographic variation in patterns of discrimination among types of neighborhoods. It was discovered that Hispanic auditors were provided systematically with less information than were Anglos in some neighborhoods, particularly

TABLE 8.5
Selected Indicators of Discrimination in Housing Availability, as Encountered by Minorities in Grand Junction, Colorado: 1983

Availability Indicators	*Black Auditors*	*Anglo Auditors*	*Difference*
Blacks			
Overall average number of units volunteered as serious possibilities	1.29*	1.56*	–0.27*
Overall average number of units invited to inspect	1.33	1.37	–0.04
Overall average number of units actually inspected	1.14	1.24	–0.10
Hispanics	*Hispanic Auditors*	*Anglo Auditors*	*Difference*
Overall average number of units volunteered as serious possibilities	1.32	1.29	+0.03
Overall average number of units invited to inspect	1.23	1.19	+0.04
Overall average number of units actually inspected	1.16	1.00	+0.16
Native Americans	*Native American Auditors*	*Anglo Auditors*	*Difference*
Overall average number of units volunteered as serious possibilities	1.21*	1.71*	–0.50*
Overall average number of units invited to inspect	1.21	1.14	+0.07
Overall average number of units actually inspected	1.14	1.07	+0.07

SOURCE: Center for Public-Private Sector Cooperation Audit, 1983.
*Statistically significant, .10 level, one-tailed test.

those with substantial Hispanic populations.

Specifically, in Hispanic neighborhoods, agents for rental housing

—commonly concealed the availability of units to Hispanic auditors;
—were less likely to identify any units that might meet the needs of the Hispanic auditors;
—failed to offer Hispanics a place on waiting lists; and
—invited significantly fewer Hispanics to inspect units than they did Anglos.

By contrast, most of the indicators suggest that agents in Anglo neighborhoods offered much the same basic information on rental housing availability to Hispanic and Anglo auditors. It is difficult to account for this relatively large measured discrimination against Hispanics in Denver's Hispanic neighborhoods (James & Tynan, 1986; James et al., 1984). The apparently low overall level of discrimination in Grand Junction could conceal neighborhood concentrations of discrimination as well.

Two other potential measures of discrimination against minorities are presented in Table 8.5—average numbers of housing units that auditors were invited to inspect and average numbers actually inspected. Auditors were instructed to seek to inspect at least one unit. As can be seen, virtually all auditors were successful in inspecting at least one unit. By contrast, the Denver audit found significant gaps between minority and Anglo auditors in terms of the numbers of units they were invited to inspect and the number actually inspected (James et al., 1984). It is likely that the very soft rental housing market in Grand Junction accounts for the relatively easy availability of units for minority inspection.

HOUSING TERMS AND CONDITIONS

Differential rental requirements may act as a barrier to the rental of a particular unit. Agents can define different terms and conditions for minorities and for Anglos with little chance of detection. Few indicators of terms and conditions show marked discrimination in Grand Junction. For example, minority auditors were quoted much the same monthly rent as were Anglos. Much the same security deposit requirements were also reported to both groups. Neither minority nor Anglo auditors were required to file a written application. Additionally, application fees were not required from minority or Anglo auditors. However, the soft market conditions in that city were reflected in agents' willingness to negotiate special deals with the auditors. Disproportionately, these special deals in rents or other terms were offered to the Anglo auditors, not the minorities.

In tests of discrimination against blacks, several agents stated that they would reduce the advertised rent if the auditor would rent immediately, that the stated security deposit could be waived if the auditor was serious about renting, that the auditor could enter into a lease/purchase agreement, and that the advertised house could be sold and financing could be arranged. These types of terms and conditions offered only to the Anglo auditors give them advantages over their black counterparts. In the Hispanic test, Anglo but not Hispanic auditors were offered a good price if they would buy the house instead of renting it and they were also offered a chance to negotiate the monthly rent and the security deposit. Unlike the Anglo auditors, Native Americans auditors did not receive the special advantageous offers of negotiated rents and security deposits on the advertised unit.

It is suggestive that agents requested more information on the characteristics of minority applicants—especially Hispanics—than they did of Anglos. For example, 42% of Hispanic testers were asked about their employment; only 26% of Anglos were asked about employment in these audits. Disparities in requests for information may signal disparities in tenant qualifications required by an agent. This cannot be determined, unfortunately, as audits can measure only discrimination in early stages of the process of finding housing.

NEIGHBORHOOD STEERING AND SALESMANSHIP IN GRAND JUNCTION

"Steering" and salesmanship consist of a variety of practices that have the effect of encouraging minorities to focus their housing search in neighborhoods or types of housing in which the minorities are deemed "acceptable." Anglos too may be steered away from racially or ethnically integrated environments. Thus steering in the rental housing market prevents equal access and narrows minority choices. Clear instances of steering were encountered in Grand Junction.

Case III

Ms. X, a Native American female, telephoned to make an appointment to see a unit advertised in the newspaper for rent. After talking for a while about the unit, the agent stated that this particular unit is located in a "rough" area and she may not be interested. When the auditor appeared in person, the agent seemed somewhat surprised but said that she would fit in quite well in this neighborhood. Mrs. Z, the Anglo female auditor, was indeed told that the area is "rough," was shown the unit, but was also given an alternative that she should inspect. The auditor reported that the alternative was located in a "white middle-class area."

In another instance of steering (Case I described above), a Hispanic auditor was steered to an integrated, low-income area and the Anglo team member was not. In one test of discrimination against Native Americans, the Anglo auditor was told by the agent that "the public schools weren't very good in the area because there were too many minorities." In another instance, an agent told a black auditor: "You'll like this area because there are other blacks around." To the Anglo team member, the agent stated, "I guess you know that this is an integrated neighborhood. Chicanos and blacks are around here." By contrast, the Denver audit found virtually no such comments.

DISCRIMINATION IN GREELEY

The Greeley audits were conducted in November and December 1983. As in Grand Junction, the local newspaper was used on a daily basis to draw the sample population. In Greeley, 91.6% of the agents seen by auditors were Anglo, 0.5% were black, 5.8% were Hispanic, 1.6% were Native Americans, and 0.5% were listed as "other." Signs of discriminatory concealing of information on available housing are as strong in Greeley as in Grand Junction, especially for blacks.

HOUSING AVAILABILITY

In Greeley as in Grand Junction, the audits establish that real estate agents report more units as available to Anglos than to minorities. Five black auditors were told that the advertised unit was not available for immediate inspection. Only one Anglo was told this. The following case gives a brief account of differential information given to a black and Anglo team member in Greeley.

Case IV

Mr. A, a black auditor, contacted the agent by phone to set up an appointment to see the advertised unit. The call was made at 11:00 a.m. and the appointment was set up for 3:15 the same day. The agent gave all of the particulars about the unit over the phone that coincided with what was in the newspaper. When the auditor arrived for the appointment, the agent stated that the unit had been rented and there was no chance of any other unit being available in the near future. The auditor reported that the agent was very unfriendly and at times very rude.

The Anglo auditor also called to make an appointment for the following day at 10:30 a.m. He talked to the same agent as did the black auditor. The agent stated that the advertised unit was available for immediate inspection and that the neighborhood surrounding the unit was quiet and had nice people living there. He stated that "he had to be very selective in renting." The agent not only showed the auditor the advertised unit, but other units that were available and in another building. He also suggested that if the auditor were really interested, they could strike a deal with reference to lowering the security deposit.

Like blacks, Hispanic auditors in Greeley were provided with inferior information on their housing options, compared to the information provided to Anglos. The following cases give an account of what took place when the Hispanic and Anglo auditors tried to rent the advertised unit.

Case V

Ms. C, the Hispanic auditor, appeared at the complex at 9:30 a.m. She was told that the unit was not available for inspection but would be in a few days. The auditor reported that the agent was very friendly throughout the interview, asking about her educational background, where she worked previously, and mentioned that "there were some Spanish-speaking Mexicans down the block." The agent suggested an alternative unit that was very nice and stated that the auditor would "love it there." The auditor drove by the address that was given to her by the agent and found that the unit was in a dilapidated trailer park.

Mrs. D, the Anglo auditor, appeared one hour later and was shown the advertised apartment. The agent took her name and telephone number.

Case VI

Mr. X, a Hispanic auditor, responded to an ad in the newspaper relative to a duplex unit. The unit was dirty and badly in need of repairs. The agent offered to negotiate the rent and the deposit if the auditor were interested.

Mr. Y, an Anglo auditor, responded to the same ad. The agent told the auditor that this was not the kind of neighborhood that he would recommend. He stated that the neighborhood had changed and the "Mexican gangs from Brighton were always hanging around." The agent suggested three alternatives in different areas of the city. The auditor reported that the areas were white, middle class.

Table 8.6 presents basic quantitative indicators of the information provided minority and Anglo auditors in Greeley. As can be seen, black auditors were told of 25% fewer serious possibilities than were Anglos in these tests. They were also invited to inspect fewer units, and actually inspected fewer units. More strikingly, approximately 20% of black auditors were volunteered no units as serious possibilities, while no

TABLE 8.6
Selected Indicators of Discrimination in Housing Availability, as Encountered by Minorities in Greeley, Colorado: 1983

Availability Indicators	*Black Auditors*	*Anglo Auditors*	*Difference*
Blacks			
Overall average number of units volunteered as serious possibilities	1.04**	1.38**	–0.34**
Overall average number of units invited to inspect	1.04	1.21	–0.17
Overall average number of units actually inspected	0.85	1.00	–0.15
	Hispanic Auditors	*Anglo Auditors*	*Difference*
Hispanics			
Overall average number of units volunteered as serious possibilities	1.00	1.14	–0.14
Overall average number of units invited to inspect	0.94	0.97	–0.03
Overall average number of units actually inspected	0.91	1.03	–0.12
	Native American Aduitors	*Anglo Auditors*	*Difference*
Native Americans			
Overall average number of units volunteered as serious possibilities	1.16	1.13	+0.03
Overall average number of units invited to inspect	1.06	0.88	+0.18
Overall average number of units actually inspected	1.03	0.82	+0.21

SOURCE: Center for Public-Private Sector Cooperation Audit, 1983.
**Statistically significant, 10% level, one-tailed test.

Anglos fell into this category. Of the black auditors, 20% were not invited to inspect any unit, while only 4% of the Anglo auditors fell into this category. In 24% of the cases, blacks were not able to inspect any unit as opposed to 8% of the Anglos.

Quantitative evidence for discrimination is apparent but weaker for Hispanics. Hispanic auditors were told of 12% fewer serious possibilities than were Anglos, and actually inspected 12% fewer units. Surprisingly, the evidence suggests that Native Americans sometimes receive *superior* information, compared to Anglos. There is no evidence of systematic discrimination in Greeley against Native Americans in terms of housing availability. The relatively little apparent discrimination against Native Americans may be the result of the high average economic status of Native Americans in Greeley. As has been pointed out, the median household income of Native Americans in 1980 was much higher than that of whites. Nonetheless, Native Americans did encounter some discrimination, as the following case makes clear.

Case VII

Ms. Y, a Native American, received all of the particulars on the telephone. The agent sounded eager to rent the unit. When the auditor arrived, the agent appeared somewhat surprised. The whole story changed. The agent stated that the single-family house was available, but "the owners were having second thoughts about renting the house." In the event they decided to rent, he would give the auditor a call.

The same agent offered the single-family dwelling to the Anglo auditor.

TERMS AND CONDITIONS IN GREELEY'S RENTAL MARKET

A majority of the terms and conditions indicators suggest that differential terms and conditions were set for black auditors as opposed to Anglo auditors in Greeley. Rental agents suggested that there were no lease requirements for 14.3% of the black auditors, while 39.1% of Anglo auditors were told that no lease was required; 85.7% of the blacks were required to sign a lease of up to one year, while 60.9% of Anglos were required to sign a lease for the same period.

More strikingly, the average monthly rental quoted to black auditors was higher than that of Anglo auditors. The amount quoted to blacks was $241.00 as opposed to $214.00 quoted to Anglos. Quoted security deposits often differed between black auditors and Anglo auditors. In some instances, as in Grand Junction, Anglos were told about "specials" that reduced the amount of deposit required.

The evidence for differential housing terms is clear, but once again, weaker for Hispanics. In some of the Hispanic tests, agents offered to reduce the security deposits if the Anglo auditors would move in immediately. Generally, such offers were not made to the Hispanic auditors. Agents also requested markedly more information from Hispanic auditors than they did from Anglos. For example, 63% of Hispanic auditors were asked about their employment. Only 31% of

Anglos were. Similarly, one-third of the Hispanic auditors were required to provide references; only 14% of the Anglos were.

Even weaker evidence was found to suggest discrimination against Native Americans. Agents were aggressive in inviting Native Americans to apply for rental housing. The agents invited 15.2% of Native American auditors to file an application as opposed to 7.4% of the Anglo auditors. However, 9.5% of the Native Americans were asked for an application fee, while none of the Anglos were required to file an application fee. In addition, 24% of the Native Americans were asked to submit to credit checks. Only 6% of Anglos were asked to do the same.

NEIGHBORHOOD STEERING IN GREELEY'S RENTAL MARKET

Unlike Grand Junction, the audits provide no evidence that steering is taking place in Greeley. In one case, an Anglo auditor was told that "quite a few blacks have been moving into the complex recently." Agents made no references to Anglo auditors about Hispanics or other minorities, but did make comments to four Hispanic auditors, that is, "you'll like it here because there are other Mexicans around," and "all of the Chicanos generally pay their rent on time." Agents made comments about Native Americans to 3.7% of the Anglo auditors as opposed to 13.8% of the Native American auditors, that is, "we all like Native Americans."

CONCLUSIONS

Minorities—blacks, Hispanics, and Native Americans—were subject to obvious forms of discrimination by rental agents or landlords in both Grand Junction and Greeley. All three groups were given fewer options than their Anglo counterparts, thus hampering their ability to make a choice. Agents made little or no attempt to negotiate rents, security deposits, or other special considerations with minority auditors. By contrast, a variety of special deals were offered to Anglos in both Grand Junction and Greeley. While steering is difficult to detect in this study since the minority communities of the cities are so small, there were apparent examples.

Clearly, both the Grand Junction study and the Greeley study point to the fact that access to rental housing remains unequal for minority groups. Blacks encountered more discrimination on the availability and terms and conditions indicators in Greeley, Colorado, than did Hispanics or Native Americans. In Grand Junction, Native Americans experienced more discrimination than did blacks or Hispanics.

MINORITY HOUSING CONDITIONS IN COLORADO'S NONMETROPOLITAN COMMUNITIES

Minority-Anglo disparities in housing conditions can indicate the extent to which minority housing choice is constrained by discrimination. Two specific indicators of discrimination—differences in the probability of home ownership and differences in dwelling size—are examined here. Differences between Hispanics and Anglos are analyzed for a sample of households in nonmetropolitan Colorado. In addition, statistical analyses of black-Anglo and Hispanic-Anglo differenccs are conducted for a sample of households in the Denver metropolitan area. This allows a comparison of the extent of discrimination faced by Hispanics living in nonmetropolitan areas with the extent of discrimination experienced by metropolitan Hispanics and blacks.

One indicator of discrimination is the existence of differences in the probability of home ownership among races and ethnic groups. In the absence of discrimination, one would expect a black or Hispanic household to have the same probability of owning a home as an Anglo household with the same income and demographic characteristics. Several studies have found that blacks in metropolitan areas have a much lower probability of home ownership than do similar whites (see, for example, Bianchi, Farley, & Spain, 1982; Kain & Quigley, 1972). Although the focus in the literature has been on discrimination encountered by blacks, a recent study shows that metropolitan Hispanics as well as blacks have a lower probability of home ownership than do Anglos (James et al., 1984). Another indicator of discrimination is a finding of differences in dwelling-unit size, after controlling for income, household size, and other demographic characteristics. That discrimination can produce differences in dwelling size by race and ethnicity is suggested by additional results of the James and Tynan study showing that units occupied by Hispanics and blacks are more likely to be overcrowded than units occupied by similar whites.

Minority-Anglo differences in the probability of owning a home and differences in dwelling-unit size are analyzed here for two 500-household samples from a database of Colorado households obtained from the 1980 census Public-Use Microdata Samples for Colorado. These include a sample of Hispanic and Anglo households in nonmetropolitan Colorado and a sample of black, Hispanic and Anglo households in the Denver metropolitan area. The small number of black households in nonmetropolitan areas of the state necessitated their exclusion from the nonmetro sample.

MINORITY-ANGLO DIFFERENCES IN THE PROBABILITY OF HOME OWNERSHIP

In the case of home ownership, the statistical analysis involves a binary dependent variable (1 = own, 0 = rent). A logit model of the relationship among the probability of home ownership and several explanatory variables is used. The explanatory variables include a variable measuring household income and dummy variables that group households according to age and household type. The model is specified as follows:

$$\log\left(\frac{P_i}{1-P_i}\right) = \alpha + \beta_1 Y + \beta_2 H_1 + \beta_3 H_2 + \beta_4 H_3 + \beta_5 H_4 + \beta_6 H_5 + \beta_7 H_6 + \beta_8 H_7 + \beta_9 R + \beta_{10} E$$

where

P_i = probability that household owns its home

Y = household income in dollars

H_1 = 1 if married-couple family with a householder under 45 years and 0 otherwise

H_2 = 1 if married-couple family with a householder over 45 years and 0 otherwise

H_3 = 1 if family with a female householder, no husband present, under 45 years and 0 otherwise

H_4 = 1 if family with a female householder, no husband present, over 45 years and 0 otherwise

H_5 = 1 if nonfamily household with a male householder under 45 years and 0 otherwise[10]

H_6 = 1 if nonfamily household with a male householder over 45 years and 0 otherwise

H_7 = 1 if nonfamily household with a female householder over 45 years and 0 otherwise

R = 1 if householder is black and 0 otherwise (metro sample only)

E = 1 if householder is Hispanic and 0 otherwise

Nonfamily households with a female householder under age 45 are incorporated in the constant term. Hispanics are defined as nonblacks

TABLE 8.7
Maximum-Likelihood Estimates:
Probability of Homeownership[a]

Independent Variable	*Metro Sample*		*Nonmetro Sample*	
Household Income ($)	7.36x10^{-5}	(6.30)*	4.95x10^{-5}	(4.44)*
Married-couple family with householder under age 45	1.62	(3.94)*	2.06	(3.44)*
Married-couple family with householder over age 45	2.73	(5.65)*	2.83	(4.62)*
Family with female householder under age 45	1.26	(2.21)*	–.321	(–.366)
Family with female householder over age 45	1.38	(1.81)*	3.06	(3.15)*
Nonfamily male householder under age 45	–.159	(–.332)*	.193	(.294)
Nonfamily male householder over age 45	2.25	(3.23)*	2.73	(3.68)*
Nonfamily female householder over age 45	1.88	(3.82)*	2.00	(3.19)*
Black householder	–.683	(–1.32)**		
Hispanic householder	–.673	(–1.78)*	–.545	(–1.67)*
Constant	–2.22	(–5.53)*	–1.95	(–3.42)*
Number of observations	500		500	

NOTE: T-statistics are in parentheses.
a. In the case of a multivariate logit model, there is no simple interpretation of the magnitude of the coefficients. Interpretation of the magnitude of the coefficients on the Hispanic and Black dummy variables requires calculation of Hispanic-Anglo and Black-Anglo differences in the probability of home ownership from the estimated equations, for different values of the independent variables (see Table 8.8).
*Statistically significant, .05 level; **Statistically significant, .10 level.

of Spanish origin. Anglo households are defined as white households not of Spanish origin.

The model is estimated using a maximum-likelihood nonlinear estimation procedure. The estimation results for both the metro and nonmetro sample are shown in Table 8.7. Both of the estimated equations converge and are significant at the .05 level using a likelihood ratio test. With the exception of the coefficients on the dummy variables for nonfamily households with a male householder under 45 (for both equations) and the dummy variable for families with a female householder under 45 in the nonmetro equation, all of the coefficients are significant (at the 0.5 level in all but one case). The coefficients indicate that the probability of home ownership is positively related to household income, as expected, and that compared to nonfamily households with a female householder under 45, all of the other age-household-type groupings with significant coefficients have a higher probability of home ownership. Households with older heads have a higher probability of owning a home than households of the same type with younger heads.

Results indicate that blacks and Hispanics have a significantly lower probability of home ownership than do similar Anglos. For the metro sample, the coefficients on the dummy variables for both black and Hispanic are negative. The coefficient for blacks is significant at the .10 level; for Hispanics, it is significant at the .05 level. For the nonmetro sample, the coefficient on the Hispanic variable is negative and significant at the .05 level. When a multivariate logit model is used, there is no simple interpretation of the magnitude of the coefficients. Interpreting the magnitude of the coefficients on the dummy variables for minorities requires calculation of differences in the probability of home ownership from the estimated equations, for different values of the other independent variables.

Table 8.8 shows racial and ethnic differences in the probability of home ownership for different age-household-type groups. For each group, the figures show the difference in the probability of owning a home for a household with the mean income for the group. For the metro sample, the magnitude of black-Anglo and Hispanic-Anglo differences are very similar. The smallest differences are found for married-couple families with a householder over age 45; the probability that a black or Hispanic household with mean income for this group owns a home is 5.0 percentage points lower than the probability than a similar Anglo household will own a home. The probability for female-headed families with a household under 45 and nonfamily female householders over 45 is 16.6 percentage points lower for Hispanics (and 16.8 percentage points lower for blacks) than for similar Anglo households.

Hispanic-Anglo differences in the probability of home ownership for the nonmetropolitan sample are somewhat smaller in magnitude than the corresponding differences for the metro sample. Results indicate, however, that Hispanics living in nonmetropolitan areas, as well as Hispanics living in metropolitan areas, have a substantially lower probability of owning a home than do Anglos. In the case of married-couple families with householders over age 45, the Hispanic-Anglo difference in the probability of home ownership is larger for the nonmetro sample than for the metro sample.

TABLE 8.8
Minority-Anglo Differences in the Probability of Home Ownership for Household Groups with Mean Income for Group

Household Group	*Metro Sample*		*Nonmetro Sample*
	Hispanic	*Black*	*Hispanic*
Married-couple families with householder under 45	–.130[1]	–.133	–.109
Married-couple families with householder over 45	–.050	–.050	–.072
Family with female householder under 45	–.166	–.168	–.057
Family with female householder over 45	–.165	–.167	–.086
Nonfamily male householder under 45	–.100	–.102	–.095
Nonfamily male householder over 45	–.158	–.161	–.111
Nonfamily female householder under 45	–.096	–.097	–.067
Nonfamily female householder over 45	–.166	–.168	–.135

1. A negative sign indicates that the group has a lower probability of homeownership than do Anglos.

MINORITY-ANGLO DIFFERENCES IN DWELLING SIZE

Racial and ethnic differences in dwelling-unit size are estimated using ordinary least squares. The dependent variable used as a measure of size is a continuous variable measuring the number of rooms in the housing unit. Explanatory variables include household income, household size, and age-household-type dummy variables. The following model is specified:

$$S = \alpha + \beta_1 P + \beta_2 Y + \beta_3 H_1 + \beta_4 H_2 + \beta_5 H_3 + \beta_6 H_4 + \beta_7 H_5 + \beta_8 H_6 + \beta_9 H_7 + \beta_{10} R + \beta_{11} E$$

where

S = number of rooms in housing unit

P = number of persons in household

Y = household income in dollars

H_1 = 1 if married-couple family with a householder under 45 years and 0 otherwise

H_2 = 1 if married-couple family with a householder over 45 years and 0 otherwise

H_3 = 1 if family, with a female householder, no husband present, under 45 years and 0 otherwise

H_4 = 1 if family with a female householder, no husband present, over 45 years and 0 otherwise

H_5 = 1 if nonfamily household with a male householder under 45 years and 0 otherwise

H_6 = 1 if nonfamily household with a male householder over 45 years and 0 otherwise

H_7 = 1 if nonfamily household with a female householder over 45 years and 0 otherwise

R = 1 if householder is black and 0 otherwise (metro sample only)

E = 1 if householder is Hispanic and 0 otherwise

Again, nonfamily households with a female householder under age 45 are included in the constant term.

Estimation results for both the metro and nonmetro samples are shown in Table 8.9. With the exception of the coefficients on dummy variables for nonfamily households with male householders, all of the coefficients on the variables measuring income and demographic characteristics are significant, in all but two cases at the .05 level. The number of rooms in a unit is positively related to household income and household size. Compared with nonfamily households with a female householder under age 45, most of the age-household-type groupings occupy units with a significantly larger number of rooms.

The metro sample results indicate that Hispanics occupy units with a significantly smaller number of rooms, .76, than do comparable Anglos. Hispanics in nonmetropolitan Colorado occupy units with .23 fewer rooms than units occupied by similar Anglos; this difference is not statistically significant. The smaller Hispanic-Anglo difference found for the nonmetro sample may reflect differences in the housing stock between the two types of areas. It may be the case that a lower cost of space in nonmetropolitan areas results in less variation in the housing stock by number of rooms. A very small and statistically insignificant

TABLE 8.9
OLS Estimates: Dwelling Size

Independent Variable	*Metro Sample*		*Nonmetro Sample*	
Household Size	.404	(6.03)*	.030	(1.46)**
Household Income ($)	4.56×10^{-5}	(9.53)*	4.33×10^{-5}	(8.16)*
Married-couple family with householder under 45	.842	(2.85)*	1.34	(3.63)*
Married-couple family with householder over 45	1.04	(3.54)*	1.57	(4.24)*
Family with female householder under age 45	.892	(2.21)*	.959	(1.93)*
Family with female householder over age 45	1.43	(2.58)*	1.38	(2.36)*
Nonfamily male householder under age 45	.036	(.120)	–.175	(–.430)
Nonfamily male householder over age 45	.602	(1.25)	.322	(.683)
Nonfamily female householder over age 45	1.06	(3.15)*	.643	(1.61)**
Black householder	–.003	(–.009)		
Hispanic householder	–.760	(–2.82)*	–.233	(–1.03)
Constant	2.91	(11.5)*	3.37	(9.87)*
Number of observations	500		500	
R-squared	.401		.298	
Adjusted R-squared	.387		.283	

*Statistically significant, .05 level; **Statistically significant, .10 level.

difference is found between the number of rooms in units occupied by blacks in Denver and units occupied by Anglos with the same income, household size, and in the same age-household-type grouping.

In summary, the analyses suggest that the extent to which discrimination constrains housing choice, as indicated by differences in the probability of owning a home, is similar for metropolitan blacks, metropolitan Hispanics, and nonmetropolitan Hispanics. All three groups have a substantially lower probability of home ownership than do comparable Anglos. In addition, Hispanics in both metropolitan and nonmetropolitan areas live in units with a smaller number of rooms than do Anglos with similar socioeconomic characteristics, although the Hispanic-Anglo difference is smaller in nonmetropolitan areas.

CONCLUSIONS

The evidence unambiguously suggests that blacks, Hispanics, and Native Americans encounter marked housing discrimination in Colorado's small cities and nonmetropolitan communities. It is also clear that this discrimination circumscribes their housing opportunities, particularly their access to home ownership. These findings are to be expected. Minorities are more vulnerable to discrimination in such places, because effective collusion to limit minority housing choice is more readily achievable in smaller communities than in major metropolitan places such as Denver. In addition, competition among rental and sales agents, mortgage lenders, and insurers is likely to be less stringent in such communities.

These findings emphasize the importance of effective enforcement of fair housing legislation in small cities and rural settings, as well as in large cities. The effectiveness of fair housing enforcement efforts is questionable under the best of circumstances. Some evidence suggests that such laws are especially impotent in small cities and rural settings. One type of evidence is the frequency of complaints to fair housing agencies regarding discrimination. Colorado, for example, has a state law, enforced by the Colorado Civil Rights Division, which bans most types of housing discrimination. Division staff report that 90% to 95% of fair housing complaints are filed by residents of the Denver metropolitan area, even though half of Colorado's populations lives *outside* this metro area (James et al., 1984, p. 126).

Additional evidence of the impotence of fair housing legislation comes from a recent evaluation of HUD's enforcement of Title VIII, the principal national fair housing law (Wallace & Lane, 1985). HUD receives, investigates, and processes fair housing complaints in its ten regional offices, located in the major urban centers of the United States. The evaluation concluded: "In remote geographic areas, intake was slowed while HUD staff attempted to secure notarized complaints" (Wallace & Lane, 1985, p. 26). This same study foeused special attention on four regional offices. It went on to point out that at all four regional offices:

> An Intake Specialist assisted potential complainants in completing the 903 Complaint Form and then notarized the Form. In the remote locations of one Region, community fair housing groups assist with this process, by reviewing jurisdiction (ascertaining that filing is within the appropriate time limit, that the complainant is a member of a protected group and that the dwelling in question in covered by Title VIII) and

> assisting complainants to draft written complaints. The Regional office's staff felt that this assistance was an important technique for assuring service to locations that their staff could not easily reach and felt that additional funds should become available to expand the practice. The 903 form was still completed and notarized at HUD. (Wallace & Lane, 1985, p. 16)

It is to be expected that many potential complainants from nonmetropolitan areas or small cities are discouraged by the greater difficulty of filing a complaint from such areas.

The audit results from Grand Junction and Greeley provide a third type of evidence. The brazenness of some instances of discrimination in these cities strongly suggest a substantial freedom of agents from fear of detection and punishment by fair housing agencies.

Clearly, then, steps should be taken to strengthen the enforcement of fair housing laws in small cities and nonmetropolitan areas. Easing complaint procedures in these communities will not alone be enough. The Colorado Civil Rights Division has branch offices in both Grand Junction and Greeley. At a minimum, both minorities and leaders in such communities must be convinced that such fair housing agencies are present in such places, that they mean business, and that they can help.

APPENDIX

AUDIT PROCEDURES AND AUDIT FORMS

The study's population consisted of units offered for rent by real estate agents or apartment managers. The local newspapers in Greeley and Grand Junction were used to identify advertised rental units. A sample was drawn from the lists of advertised units; audit assignments were made on a daily basis. Teams of auditors were matched (one black and one Anglo; one Hispanic and one Anglo; one Native American and one Anglo) according to characteristics such as age, income, sex, family size, occupation, and marital status. The first auditor contact with agents or landlords was to be made on the same day of sampling; the second team member was instructed to follow 2-4 hours later.

A total of 16 auditors were utilized in the Grand Junction study—5 blacks, 3 Hispanics, 3 Native Americans, and 5 Anglos (10 females and 6 males). A total of 16 auditors were utilized in the Greeley study—4 blacks, 4 Hispanics, 3 Native Americans, and 5 Anglos (10 females and 6 males). Audit reporting procedures and forms were essentially identical to those used in the earlier Denver audit (James et al., 1984).

NOTES

1. The audit technique used to measure discrimination presents difficulties in very small cities or in rural communities where the real estate industry is tight-knit and there is widespread familiarity with the people of the area. Indeed, an effort in 1983 to perform a similar audit in Durango, Colorado, failed when the audit was discovered by local officials and the audit team refused to work on the project any longer. The failure of the Durango audit in the summer of 1983 led us to initiate the Greeley audit that fall.

2. Whites include both Caucasian Hispanics and Anglos.

3. These measures are not strictly comparable to those presented in the table.

4. These are tracts 6 and 7.01.

5. These are tracts, 1, 5, 6, 7.01, 7.02, and 13.

6. These are tracts 2, 3, 8, and 10.

7. This is tract 9.

8. The audit project was directed by Franklin James. The field work was managed by Josephina Vilar. Initial analyses of the audit findings were presented in McCummings, 1984.

9. "Serious possibilities" were the advertised unit, if described as available, plus available "comparable" units, plus other available units described as similar to the advertised unit.

10. A nonfamily household is a one-person household or a household composed of two or more unrelated individuals.

REFERENCES

Bianchi, S. M., Farley, R., & Spain, D. (1982). Racial inequalities in housing: An examination of recent trends. *Demography.*

Colorado Civil Rights Division. (1981). *Hearing report of the Western Slope Study.* Denver, CO: Author.

Feins, J. D., & Bratt, R. G. (1983, Summer). Barred in Boston: Racial discrimination in housing. *Journal of the American Planning Association, 49*, 344-355.

Feins, J. D., Bratt, R. G., & Hollister, R. (1981). *Final report of a study of racial discrimination in the Boston housing market.* Cambridge, MA: Abt.

Feins, J. D., & Holshouser, W. L., Jr. (1984). *The multiple uses of audit-based research: Evidence from Boston.* Paper presented at the HUD conference on Fair Housing Testing.

Grebler, L., Moore, J., & Guzman, R. (1970). *The Mexican American people: The nation's second largest minority.* New York: Free Press.

Hakken, J. (1983). *Housing the Hispanic population: Are special programs and policies needed?* Unpublished report, Office of Policy Development and Research, Washington, DC: Department of Housing and Urban Development.

James, F. J., McCummings, B., & Tynan, E. A. (1984). *Minorities in the sunbelt: Segregation, discrimination and housing conditions of Hispanics and blacks.* New Brunswick, NJ: Rutgers University Center for Urban Policy Research.

James, F. J., & Tynan, E. A. (1986). Segregation and discrimination of Hispanic Americans: An exploratory analysis. In J. Goering (Ed.), *Desegregation, race and public policy.* Chapel Hill, NC: University of North Carolina Press.

Kain, J. F. (1980). *National urban policy paper on the impacts of housing market discrimination and segregation on the welfare of minorities.* Cambridge, MA: Harvard University.

Kain, J. F., & Quigley, J. M. (1972). Housing market discrimination, home ownership, and savings behavior. *American Economic Review, 62,* 263-277.

Kain, J. F., & Quigley, J. M. (1975). *Housing markets and racial discrimination.* New York: National Bureau of Economic Research.

deLeeuw, F., Schnare, A. B., & Struyk, R. J. (1976). Housing. In N. Glazer & W. Gorham (Eds.), *The urban predicament.* Washington, DC: Urban Institute.

Lieberson, S. (1980). *A piece of the pie.* Berkeley: University of California Press.

Massey, D. S., & Mullen, B. P. (1984, January). Processes of Hispanic and Black spacial assimilation. *American Journal of Sociology, 89,* 836-873.

McCummings, B. (1984). *Testing for housing discrimination against blacks, Hispanics and Native Americans in Grand Junction and Greeley, Colorado.* Denver: Colorado Civil Rights Division.

McCummings, B. (1985). *The report on the University of Northern Colorado.* Denver: Colorado Civil Rights Division.

Newburger, H. (1984, April). *Recent evidence on discrimination in housing* (HUD-PDR-786). Washington, DC: Department of Housing and Urban Development.

Orfield, G. (1980). *Housing and school integration in three metropolitan areas: A policy analysis of Denver, Columbus and Phoenix.* Washington, DC: Department of Housing and Urban Development.

U.S. Bureau of the Census. (1983). *Public-use microdata samples technical documentation.* Washington, DC: Department of Commerce.

U.S. Bureau of the Census. (1984). *Statistical abstract of the United States: 1985.* Washington, DC: Government Printing Office.

U.S. Department of Housing and Urban Development. (1979a, April). *Measuring racial discrimination in American housing markets: The housing market practices survey.* Washington, DC: Author.

U.S. Department of Housing and Urban Development. (1979b). *Discrimination against Chicanos in the Dallas rental housing market: An experimental extension of the housing market practices survey.* Washington, DC: Author.

Wallace, J. E., Holshouser, W. L., Lane, T.S., & Williams, (1985). *Evaluation of the fair housing assistance program.* Cambridge, MA: Abt.

Wallace, J. E., & Lane, T. S. (1985). *Case study on HUD processing of Title VIII fair housing complaints.* Cambridge, MA: Abt.

Yezer, A. (1980, July). *How well are we housed? 1. Hispanics* (HUD-PPD-33393). Washington, DC: Department of Housing and Urban Development.

Yinger, J. (1979). Prejudice and discrimination in the urban housing market. In P. Mieszkowski & M Straszheim (Eds.), *Current issues in urban economics.* Baltimore: Johns Hopkins University Press.

9

The Roots of Segregation in the Eighties: The Role of Local Government Actions

YALE RABIN

ONE OF THE MOST PERSISTENT and pervasive characteristics of U.S. metropolitan areas is residential segregation by race. While gross national data on suburbanization of blacks during the 1970s may, at first glance, appear to suggest some improvement, this optimistic view is readily dispelled by closer scrutiny. Although the black population outside the central cities of metropolitan areas increased by 2.8 million between 1970 and 1980, a 43% gain, black suburban residents are disproportionately concentrated outside the central cities of a small number of large SMSAs. Over half of all suburban blacks are in the outer rings of seven SMSAs, and their developing spatial distribution there exhibits familiar patterns of racial segregation (Rabin, 1983). And while large-scale suburbanization of blacks has occurred in these few places, the continuing increase of blacks as a proportion of central city population has been widespread (Long & DeAre, 1981). Contributing to this concentration has been a concurrent out-migration of blacks from suburbs to central cities that continued at significant levels during the 1970s (Nelson, 1979).

Kain's observation in 1974 that this process of racial segregation "has created major distortions in the patterns of metropolitan growth, and bears a major responsibility for a surprisingly long list of ills" (p. 16) is as accurate in the mid-1980s as it was then; and among these ills, segregated schools and isolation from decentralized employment opportunities remain as chronic disorders. Governments at all levels have failed to provide adequate or appropriate remedies.

This failure is attributable, in part, to unreasonably constricted views of the causes of segregation. These views ignore the critical effects of public intervention in land development in general and in housing markets in particular. A substantial body of research, focused almost exclusively on housing market behavior, supports the notion that certain demographic characteristics and racial attitudes primarily influence the behavior of buyers and sellers as they participate in the operation of housing markets. Data for these many studies have generally been derived from the census and from public opinion surveys (Streitweiser & Goodman, 1983).

But other studies show compelling evidence that disparities in income are not sufficient to explain the extent to which blacks are segregated from whites (Tauber & Tauber, 1965). This conclusion has been reinforced by more recent findings that increases in income for blacks yield increases in integration, while increases in income for whites result in increase in segregation (Jones, 1982).

In reviewing a series of attitude surveys, Pettigrew (1973) found a steady decline in expressions of resistance by whites to integrated living, but concluded, based on his own study, that "there remains an enormous degree of fear, reluctance and downright opposition" (pp. 32-33). More recently, Farley et al. (1983), in a more probing analysis, concluded that earlier studies have tended to understate the extent and degree of white opposition to integration. Another review in a position paper on the relationship between school segregation and housing segregation, which was signed by 37 social scientists, noted emphatically that "every major study of the housing of Blacks and Whites has identified racial discrimination as a major explanation of the observed segregation" (Orfield et al., 1980). Evidence that "white flight" is motivated more by hostility to blacks than opposition to busing (Cusick, Gerbing, & Rossel, 1979) serves to further corroborate these findings.

A concise, but cogent overview of this discrimination-segregation relationship is provided by Yinger (1979, p. 459) who found that the evidence

> overwhelmingly supports the proposition that racial discrimination is a powerful force in urban housing markets. Only a theory that involves

> discrimination can explain why blacks are concentrated in a central ghetto, why blacks pay more for comparable housing than whites in the same submarkets, why prices of equivalent housing are higher in the ghetto than in the white interior, and why blacks consume less housing and are much less likely to be home owners than whites with the same characteristics. This evidence of discrimination, based on recent data, makes a convincing case for government intervention in the housing market.

Given the persistent transformation of old patterns of segregation into new ones, the repeated finding that these conditions are attributable to racial discrimination by whites is hardly surprising. Nor is it surprising, since the vast majority of these studies have focused on the influence of discrimination on market behavior, that the inference is readily and widely drawn that racial discrimination in the private housing market is the only significant cause of housing segregation. However, since whites make up the overwhelming majority of those who govern, legislate, judge, administer, and enforce in this society, it does seem remarkable that little attention has been devoted to the ways in which this widespread racial prejudice may have influenced the nature and implementation of housing-related public policies and programs, and the behavior of public officials in those government agencies whose activities affect housing patterns.

To be sure, several studies have acknowledged the segregative effects of some, mainly discontinued, government practices, including the enforcement of racially restrictive covenants, racially discriminatory FHA policies, the segregative site selection and tenant assignment practices of local public housing authorities, the implementation of urban renewal programs, and the ongoing practice of exclusionary zoning (for example, Feagin & Feagin, 1978; Foley, 1973; Orfield et al., 1980). Generally, however, these government actions are seen as peripheral to the central issue of housing market discrimination; and no study was found that dealt, even in relative terms, with the extent of the impact of these government practices on existing segregation or attached any importance to their influence on emerging patterns of segregation.

My findings, derived from empirical case studies conducted over the past twenty years in over fifty cities, towns, and counties,[1] reveal that while widespread hostility to blacks may be a major influence on the kinds of locational decisions made in private housing transactions, the actual spatial distributions that result are strongly influenced by public actions. *Indeed, the land use-related policies and practices of government, at all levels, have been, and in many cases continue to be, important influences on both the creation and the perpetuation of racially segregated housing patterns.*

These relationships are most clearly evident in metropolitan areas where, since World War II, massive federal investments in highways have created new patterns of access and spawned millions of acres of new suburban development, providing new housing and job opportunities for millions of young, white, middle-income families. These new communities excluded blacks first by state-enforced racially restrictive covenants and the official segregation policies of the Federal Housing Administration, and then by locally enacted exclusionary zoning. These mutually reinforcing actions, either concurrently or successively, have provided both impetus and guidance to the dynamic process from which the landscape of metropolitan racial polarization continues to develop.

The pervasive influence of racist attitudes on the implementation of public programs is vividly illustrated by the failure of government agencies at all levels to implement the Congressionally enacted response to these metropolitan developments. The racially and economically polarizing consequences of the highway-stimulated process of metropolitan decentralization were recognized relatively early. Beginning in 1962, Congress responded with a remarkably consistent series of acts clearly expressing its concern about the decline of central cities and public transportation, and the plight of inner-city minority residents. Determination was also expressed to prevent or minimize the further adverse social, economic, and environmental impacts of highways and other federally funded programs.[2] These laws required the establishment of regional planning agencies, the formulation of regional plans, and the evaluation of local applications for federal funds for their compliance with regional plans and federal goals. They prohibited racial discrimination in the benefits of federally supported programs, and required the identification and public disclosure of the probable adverse impacts of such programs at public hearings and in environmental impact statements. However, these requirements, at least to the extent that they might have protected the welfare and expanded the housing opportunities of inner-city blacks, have been largely ignored, a response implicitly endorsed by token compliance reviews and encouraged by negligible enforcement (Rabin, 1980).

SEGREGATIVE LOCAL GOVERNMENT ACTIONS

In the dynamic interplay of public actions and private attitudes on racial segregation, federal funds and state-enabling legislation often combined to provide incentive and authority for local action. However, while the delegated powers of the states continue to provide an underlying basis for local action, severe reductions in the availability of federal funds for local land use-related purposes have significantly reduced direct federal involvement. While these intergovernmental relationships are of fundamental importance, an adequate examination of the complex web of federal-state-municipal influences on residential segregation is beyond the scope and intention of this chapter. In addition, with the probable exception of highway planning, the particular spatial outcomes of public segregative actions are most directly shaped by the decisions and initiatives of local government. Finally, some segregative actions such as exclusionary and expulsive development controls continue to be instruments of public policy at the local level. For these reasons, and to promote organizational clarity, the focus of what follows is on the segregative actions of local government.

Nevertheless, this limitation is not intended to suggest that past discriminatory policies and practices of the federal government have not been important and lasting influences on the establishment of present patterns of racial segregation, or that the recent (since 1980) shift of federal policy from inadequate civil rights enforcement to open opposition to civil rights enforcement has not substantially encouraged the current resurgence of public and private discrimination.

The principal local government actions and practices that have influenced residential location by race and that are segregative in their effect include:

—clearance and elimination of minority residential areas
—creation of physical barriers to expansion of minority areas
—exclusion of public and/or subsidized housing
—segregative relocation practices
—segregative site selection and/or tenant assignment for public and subsidized housing and schools
—exclusionary zoning
—discrimination in the provision of municipal facilities and services
—*de jure* segregation of housing or schools
—court-enforced racially restrictive covenants
—expulsive zoning

—changing or failing to change municipal or school district boundaries
—use of racial criteria identifying and planning neighborhoods
—public pronouncements by government officials that serve to reinforce racially discriminatory attitudes and practices

These actions and practices can be grouped into three general categories based on the nature of their effects on locational patterns. Some fall into more than one category:

(1) Those that eliminate areas of minority residences.
(2) Those that create barriers to the direction in which or the extent to which a minority area may expand.
(3) Those that foster the movement of minorities into minority areas or promote the transition of majority to minority areas.

The examples cited here are necessarily limited in number and descriptive detail. Those cited are not isolated or unique phenomena; they are typical of widespread practices, and have been selected because they strikingly illustrate the segregative nature of common public actions. However, in selecting examples from a large number of settings, attempting to emphasize more recent practices and grouping them by category of impact, the mutually reinforcing effects of other concurrent or sequential discriminatory and segregative actions, both public and private, is substantially understated.

BACKGROUND AND METHODOLOGY

The case studies that provided evidence of segregative public action were conducted in many regions of the country in towns as small as Osage, West Virginia and cities as large as Kansas City, Missouri and Philadelphia, Pennsylvania. Most, but not all, were undertaken in response to allegations that in the conduct of some land use-related activity, a government agency had illegally discriminated, usually against blacks but often against Hispanics and sometimes against Native Americans. These land use-related activities included locational decisions for assisted housing and schools, the planning and construction of transportation facilities, urban renewal, the relocation of displacees, the exercise of development controls, the provision of municipal facilities and services, and similar activities funded under the Community

Development Block Grant Program. The studies were often, but not always, carried out in support of class action civil rights litigation and were commissioned by both public interest legal assistance organizations and government agencies at the federal, state, and municipal levels.

The information base for these empirical studies was derived from examinations of change in the size, location, concentration, housing conditions, and other relevant demographic characteristics of the affected minority group, comparisons of these to characteristics of the majority white population, and historical reviews of relevant government actions. Data from the census and other available surveys and studies were collected, analyzed, and mapped at a fine level of geographic detail, whenever feasible, at block level. Changes over time were then evaluated in the context of actions and events recorded in public documents, and newspapers and library archives.[3]

DISPLACEMENT OF MINORITY AREAS

The term *displacement* is used to describe those public actions that contribute to the removal of minorities from an area. The result is an area that is subsequently redeveloped for majority residential occupancy or for some nonresidential use. The early role of clearance for highway construction and urban renewal has been widely recognized. Less well recognized, and as yet untested in the courts, is the use of zoning as an expulsive mechanism—that is, to create or facilitate market conditions that cause minority displacement. This practice, as will be shown below, is not new, and in the current absence of resources or political support for clearance, it provides an expedient, effective, and inexpensive means of pursuing the same ends. It is often a factor in gentrification that has become a significant cause of minority displacement (LeGates & Hartman, 1982).

The displacement of minority areas by public action has been observed to have one of three general spatial characteristics, each of which has perceptibly different consequences.

(1) Clearance of minority housing on the edge of a large area of minority concentration.
(2) Clearance of an entire enclave of minority housing from a majority area.
(3) Clearance within a large area of minority concentration for some nonresidential uses, either within a contained site as for a recreation area or along a linear path as for a highway.

The effect of clearance on the edge of an area of minority concentration has most often been to direct the expansion of the minority area in a direction away from the cleared area. This was observed in Kansas City, Missouri; Charlotte, North Carolina; Nashville, Tennessee; Jackson, Tennessee; Charlottesville, Virginia; and Norfolk, Virginia. In each case, the area cleared was immediately adjacent to the central business district (CBD) or government center. In each, the direction of black residential area expansion was established and continues today.

In Charlottesville, an interview was conducted with an elderly former superintendent of schools. He had been influential in 1936 in selecting a site at the edge of a black neighborhood for the construction of a large new high school for whites only. He frankly acknowledged that the determining factor was the view of the school board that "those people didn't belong there so close to our nice downtown." Less than twenty years later, the remainder of the black neighborhood was cleared by urban renewal. In Norfolk, the plan for the CBD explained that nearby housing of low-income blacks, which was subsequently cleared through renewal, placed the downtown center "in a setting inappropriate to its intended character." Clearance activities in Norfolk during the 1950s were so extensive that one out of every four black families was displaced to make way for highways or urban renewal (*Riddick v. School Board of the City of Norfolk*, 1984).

The elimination of minority enclaves appears to be among the most persistent in its impact on present neighborhood patterns. The term *minority enclave* is used to describe relatively small concentrations of minority housing ranging in size from one to perhaps ten city blocks and spatially separated from the principal minority housing area. Several such enclaves were to be commonly found in the central cities of many metropolitan areas as recently as the early 1960s, and many still exist in the outer rings. In most cities that were studied, areas from which blacks were removed thirty years ago and more recently, have no black residents in them today. The elimination of minority enclaves has been brought about by several common types of government activities, primarily clearance and displacement, but sometimes discriminatory pressure. Sometimes the results are extreme. In Osage, West Virginia, blacks made up about one-third of the population in 1960. Acquisition of the right-of-way for Interstate 79, running north-south through the town, eliminated every black-occupied dwelling in the town.

In Kansas City, three radial elements of the Interstate Highway System entering the city from the east, north, and west followed curiously winding routes, each of which eliminated a black neighborhood enclave. The effect of this clearance during the 1950s and 1960s and concurrent urban renewal displacements was to increase the proportion of the city's black population living in the principal ghetto area from 81% to 94%.

The black residents of a six-block enclave in the Whitman neighborhood in southeast Philadelphia, which was cleared in 1959 to provide a site for a public housing project, moved into available housing nearby in the predominantly white area adjacent to the cleared site. The city then sought and obtained federal approval for the designation of the entire Whitman neighborhood as a spot clearance and rehabilitation urban renewal project. This project involved the clearance of about 3% of the Whitman area housing units (about 100 dwelling units), and removed every house occupied by a black household in the blocks surrounding the public housing site (*Residents Advisory Board v. Rizzo*, 1976).

Following completion of the urban renewal project, the Whitman neighborhood association, encouraged by then mayoral candidate and later Mayor Frank Rizzo, mounted a vociferous and often violent campaign of opposition to the planned construction of the public housing project on the grounds that it would bring blacks into an all-white neighborhood. After more than 20 years of conflict, the project was ordered built by the federal courts and has now restored some integration to the Whitman neighborhood. As was the case in Kansas City and Norfolk, this period of renewal activity was characterized by the extensive growth of the major ghetto areas in north and west Philadelphia.

During the mid-1960s in Easton, Pennsylvania, there were three predominantly black enclaves north of the Lehigh River and a somewhat larger black community south of the Lehigh River. A series of urban renewal projects, which cleared the three northern enclaves, and the timely construction of subsidized housing in the majority black area south of the river combined to form the more rigid pattern of a single ghetto.

Sometimes housing available to blacks was so limited that local actions forced minority displacees to move to another municipality. In Hamtramck, Michigan during the late 1960s, the city carried out several urban renewal projects in black enclaves in order to provide land for the expansion of adjacent automobile plants. Nearly a third of the city's black population was displaced. Since the cleared areas were converted to nonresidential use, and because no relocation housing was provided, most were forced to move to adjacent Detroit.

In sections of the country that once imposed *de jure* school segregation, a powerful influence on the disappearance of minority enclaves has been the closing of the minority schools that served them. In Austin, Texas the school board closed black schools serving five black enclaves in north and west Austin, leaving black parents with the burden of transporting their children to black schools in the principal black area in southeast Austin. Within ten years four of the enclaves had disappeared.

In Mt. Laurel, New Jersey the homesites in an existing low-income black community have been designated by the zoning ordinance as nonconforming uses, thereby providing the local government with a rationale for the refusal of permits for the replacement or renovation of the housing there. A systematic process of inspections, condemnations, and demolitions is slowly but steadily eliminating the minority community there. Between 1971 and 1974, nearly a third of the households in that community were displaced; and since no affordable relocation housing was available in Mt. Laurel, they have been forced to move to other jurisdictions (*Burlington County NAACP v. Mt. Laurel*, 1974).

This form of development control, which I have termed *expulsive zoning*, has been widespread and has frequently resulted in the destabilization and elimination of minority enclaves by promoting private conversion to other uses or by creating a convenient rationale for redevelopment. The first zoning ordinances adopted during the late 1920s and early 1930s in Charlotte, Charlottesville, Kansas City, Norfolk, Jackson, Tennessee, and Selma (early zoning maps were not always available in other places) zoned only the major black areas for residential use. Separate enclaves and sometimes sections on the edge of the main black residential area were zoned for industrial or commercial uses. Although the black residential use predated the adoption of zoning, the nonconforming status of these areas formed part of the basis for later clearance through renewal in Kansas City, Jackson, and Norfolk.

In Baltimore County, Maryland, some suburban black enclaves were zoned for nonresidential uses even though adjacent white areas were zoned residential (Rabin, 1970). A recent report on displacement from El Paso's barrio revealed that population there had declined by over half in the last fifteen years, and that expulsive zoning had played a significant role. "For decades, the M-1 (light manufacturing) zoning has inflated land values, promoted commercial encroachment, and allowed chaotic land-use leaving the predominantly residential community in non-conforming status" (National Low-Income Housing Information Service, 1986).

Since the displacement through urban renewal of over 300 black families from the Old Town historic district of Alexandria in the 1970s, market pressures have increased steadily there for nineteenth-century row-houses and twentieth-century imitations. For over fifteen years, those pressures have focused on the Parker-Gray district, a long-established low- and moderate-income black residential neighborhood adjacent to Old Town whose residents fought repeated attempts by the City Council to extend the boundaries of the historic district to include their neighborhood.

In 1984, the neighborhood residents lost. In spite of an earlier finding by the Virginia Historic Landmarks Commission that the neighborhood did not merit designation as a historic area, the City designated most of the neighborhood as the Parker-Gray Old and Historic District, thus accelerating the displacement of blacks and the transformation of the neighborhood into a white upper middle-income facsimile of Old Town.

CREATION OF RACIAL BARRIERS

Public action has helped to create physical barriers that restrict minority geographic movement. The construction of highways between minority and majority residential areas is an example. Other actions influence development in ways which tend to exclude minorities. The most common physical barrier has been the limited access highway, which in numerous instances has been aligned, or was proposed to be aligned, along a route that separated minority and majority residenital areas.

In Philadelphia, the initially proposed route of the Crosstown Expressway ran east-west along a line between an affluent white area adjacent to the CBD and a low-income black area to the south. After the initial proposal was made, several years elapsed during which the process of gentrification extended the white area more than a block further to the south. It was then proposed, ostensibly on traffic grounds, that the route of the highway be shifted to the south along a line that coincidentally corresponded to the new boundary between the black and white neighborhoods. After much public opposition, the road proposal was withdrawn entirely.

In the northwestern corner of Hamtramck, the route of a highway was diverted to isolate a black enclave between the highway and an automobile plant, and the isolated area then rezoned from residential to industrial use, thereby reinforcing the barrier effect by expulsive zoning. Among other cities studied, it was observed that highways were built that formed barriers between white and minority residential areas in Charlotte, Charleston, West Virginia, El Paso, Texas, Flint, Michigan, Indianapolis, Indiana, Kansas City, and Ossining, New York. Some divisive highways did not remain effective as barriers to racial movement for more than a few years, but other publicly erected physical barriers have been more enduring.

In Jackson, Kansas City, Alexandria, and Norfolk extensive redevelopment for government and commercial uses, which displaced black residential areas, has forced the growth of the black residential areas away from central business districts. In Nashville, urban renewal land was provided to create an extension of the Music City area, which would serve as both a buffer and a barrier between the Vanderbilt University campus area to the west and the all black Edgehill neighborhood to the east. Physical barriers can result from discontinuous street systems between white and adjacent to black areas. Access to such black enclaves is usually via a single street connecting to a major artery or nonresidential street. The Catonsville area of Baltimore County provided a striking example of this isolating condition. There one could look from streets in the black neighborhood across a fifty foot wide patch of trees and underbrush to the continuation of the same streets in the adjacent white neighborhood. Interestingly, these streets were shown as continuous on the county's maps.

Lack of municipal facilities in black areas constituted barriers that kept blacks segregated. In Shaw, Mississippi, the town failed to provide the segregated section in which blacks lived with basic municipal facilities such as paved streets, sidewalks, street lights, water supply, and sewers. In appealing the ruling by the district court that the town had violated the rights of its black residents, the attorney for the town argued that it was necessary to withhold these public improvements in order to ensure that there would be an area of town in which blacks could afford to live. In Alexandria, Virginia, complaints by blacks that the city had failed to make promised neighborhood improvements under a Community Renewal Program were met with a similar response (Hammer, Siler, & George, 1976). The provision of inferior municipal facilities and services to black residential areas was also documented in Mt. Laurel and in seventeen other cities, towns, and counties in Alabama, Arkansas, Florida, Georgia, Louisiana, Mississippi, Nevada, Tennessee, and Virginia.[4]

Exclusionary zoning also forms barriers around some black enclaves in the suburbs that predate post-World War II suburbanization. The expansion of these areas is often prevented and their declines accelerated by zoning the immediately surrounding area for nonresidential use or for large-lot, low-density residential use. Several such zoning-bound communities were found in Baltimore County, Maryland (Rabin, 1970).

ACTIONS THAT PROMOTE RACIAL CONCENTRATION OR TRANSITION

The public actions that fall into this category are the most varied and have also produced some of the most persistent segregative effects. The widespread practice of concentrating public and other assisted housing in the principal minority area is prominent. Early official city planning studies carried out in Kansas City, Norfolk, and Selma used explicitly racial criteria in identifying and delineating neighborhoods. In Norfolk and Kansas City these neighborhoods continue to form the basis for current plans.

Out of a total of twenty-two family public housing projects in Norfolk, twenty-one were located in all-black residential areas and were all-black occupied. Close cooperation between the school board and the housing authority led to the practice of locating elementary schools adjacent to these all-black projects thus assuring a mutually reinforcing pattern of segregated housing and schools that persists in the 1980s. These housing projects, in 1983, accommodated one out of every four black families in Norfolk. When Philadelphia, during the late 1960s, received authorization from HUD for the acquisition of over 4,300 single-family houses under the Used House Program, the city council passed an ordinance restricting the area within which the houses could be purchased to the North Philadelphia ghetto. Philadelphia is one of the few major cities in which the Index of Segregation increased between 1970 and 1980 (Tauber & Tauber, 1983).

In Nashville, three public housing projects and one rent subsidy project, all black occupied and totaling nearly 900 units, were concentrated in a single urban renewal project within an already black residential area, thereby intensifying the levels of both racial and economic segregation. In Cuyahoga County, Ohio, the housing authority, whose jurisdiction extends over 67 municipalities had, by 1975, built all of its projects in Cleveland because none of the other jurisdictions would permit public housing to be built. Since that time under limited cooperation agreements, housing for the elderly has been built in four suburban municipalities and a total of fewer than fifty family public housing units has been produced outside Cleveland.

In Goldsboro, North Carolina, the western section of the city and a small adjacent portion of the county just west of the city are served by a majority black school district. The larger eastern section of the city and the remainder of the surrounding county are served by a majority white school district. The county housing authority, which has jurisdiction over both areas, has built public housing projects only within the boundaries of the majority black school district within the city. Kansas City and Jackson are also characterized by gross spatial disparities in the areas included within the municipal and school district boundaries, with the result that black pupils are ovewhelmingly concentrated in a single school district that serves only a portion of the city. A recent landmark ruling in a school desegregation case filed by the U.S. Department of Justice in the closing days of the Carter administration also recognized the fundamental relationship between the racial characteristics of housing occupancy and school enrollment, and found that a long history of housing-related actions by the City of Yonkers, New York has contributed significantly to maintaining segregation in both ("Judge finds," 1985). Pearce (1981) has demonstrated that racially identifiable segregated schools in situations such as these are significant in promoting the segregated locational choices of white home buyers.

The relocation of displaces from public programs has frequently had the effect of increasing the concentration in existing minority areas, as was the case in the examples from Kansas City and Easton cited above; at other times, relocation has served to bring about the transition of a majority area to a minority area. In another Kansas City example, over 1,400 black households displaced by right-of-way acquisition for a contested freeway have been relocated into a single highway department designated zone in southeast Kansas City along the path of the freeway, a process that has greatly accelerated and reinforced the ongoing transition of that zone from white to black.

In Norfolk during the 1960s, black displacees from urban renewal were referred by the redevelopment authority to two real estate firms whose offices and areas of operation were in the all-white Park Place neighborhood adjacent to the northwestern edge of the downtown ghetto. By 1970 this neighborhood was entirely black. In Jackson, Tennessee, the ongoing clearance of black residential areas west of the downtown continues to increase the concentration of blacks in east Jackson.

SUMMARY AND CONCLUSIONS

Varied and widespread actions by local government, often in the implementation of federal or state programs, have been instrumental in shaping patterns of racial segregation. Many, if not most, of these activities have resulted from the implementation of ostensibly beneficial plans and proposals by agencies such as planning departments, redevelopment authorities, housing authorities, and highway departments. These agencies, rarely, if ever, attempt to avoid segregation or actively pursue integration. While it is not possible to quantify precisely the individual or collective impacts of these actions, it is certainly reasonable to generalize about their relative effects, based in part on the numbers of households affected, the size of the areas involved, and the observable evidence of the duration of their impacts.

In some cities, such as Norfolk and Easton, the scale and scope of these segregative activities have been so great that they might reasonably be characterized as the principal determinants of racial patterns in those places. In other cities, where the scale of government segregative activity may have been smaller, the momentum and direction established by activities, such as clearance and relocation, have nevertheless had an influence far out of proportion to the numbers displaced. Displacements of large numbers of households were generally carried out over relatively short periods of time and had the effect of concentrating relocatees in a single direction or a circumscribed area. This resulted in a rapid process of racial transition, with a momentum that continues for many years. The trends thus established were then reinforced by the sites selected for assisted housing and schools, and by the erection of physical and regulatory barriers, all the while providing implicit sanction and tangible support for private discriminatory practices.

As a consequence of these actions, racial segregation continues to be one of the most deplorable conditions in urban America. The actions of the 1950s, 1960s, and 1970s, and before, have established patterns that persist today. Without active remedy, these patterns will remain. *The need for an understanding of government's role in creating and perpetuating these segregated conditions is of fundamental importance.* To suggest, as the demographic literature generally does, that preference and income levels are the only significant causes of present patterns of segregation is to reinforce a distorted view of the past, and provide a convenient rationale for the denial of public accountability and action, an excuse for the reduction of government's role in protecting civil rights, and a basis for subverting equitable outcomes in the courts.

Employing this self-serving rationale, the Reagan administration has not only reduced government's efforts on behalf of civil rights but has reversed policies that have been effective in promoting opportunities for minorities (Palmer & Sawhill, 1984), even eliminating previously required certifications that recipients of federal funds comply with civil rights laws (Palmer & Sawhill, 1984). In further pursuit of these policies, the Justice Department is seeking to dismantle ongoing efforts toward school desegregation by its intervention in a number of lawsuits on behalf of local government or public agency defendants who are seeking to relieve themselves of obligations under court-imposed desegregation plans. In some instances, the department has initiated litigation to halt the implementation of desegregation plans that had been voluntarily entered into by their participants. In the presence of well-documented, widespread racial prejudice, these policies and their racially isolating consequences can only serve to consign a significant portion of the minority population to the status of a permanent underclass, living in racially segregated areas.

As a first step, past efforts to reduce segregation and achieve a more equitable distribution of opportunities must be understood not as futile and inappropriate interference with market processes but as an unfinished task—unfinished because govenrment's commitment and effort have been grossly inadequate. Myths and misconceptions, whether promulgated by partisan politicans, or inferred from narrowly focused academic research, are fundamental obstacles to this understanding.

Segregation in the 1980s is the legacy of many decades of government complicity in the process of racial isolation. To begin to undo this systemic pattern of racial segregation will require a commitment of resources and effort on a scale that seems inconceivable in the current political climate.

At current (1986) levels, $290 billion is being spent annually—$33 million an hour—for increasingly hostile activities characterized as national defense. The diversion of just two hours per day of this expenditure would provide $24 billion per year, a sum that might conceivably fund the undertaking of a serious national effort to redress past injustices. It might be enough to initiate the long overdue process of change from institutionalized separation and inequality based on race to some approximation of the equal opportunity that the society's rhetoric proclaims. While this is not likely to happen in the near future, it will never happen unless, and until, there is widespread recognition of the role of government in creating and perpetuating the racial disparities which exist.

NOTES

1. Studies of one or more racially discriminatory practices by government agencies were carried out in Birmingham, Lowndes County, and Selma, Alabama; Pine Bluff, Arkansas; Irvine, California; Denver, Colorado; Washington, D.C.; Sanford, Florida; Ocilla, Georgia; Indianapolis, Indiana; Shreveport, Louisiana; Baltimore County, Maryland; Detroit, Flint, Hamtramck, Inkster, and Livonia, Michigan; Minneapolis and St. Paul, Minnesota; Drew, Gulfport, Itta Bena, Senatobia, Shaw, and West Point, Mississippi; Kansas City, Missouri; Las Vegas, Nevada; Camden, Mt. Laurel, and Newark, New Jersey; Huntington and Ossining, New York; Charlotte, Goldsboro, and Wilmington, North Carolina; Cuyahoga County and Defiance, Ohio; Easton and Philadelphia, Pennsylvania; Columbia, South Carolina; Chattanooga, Jackson, Knoxville, Nashville, and Pulaski, Tennessee; Austin and El Paso, Texas; Alexandria, Charlottesville, Fairfax County, Greenville County, and Norfolk, Virginia; Charleston and Osge, West Virginia.

2. This legislation included the 1962 Highway Act, 1964 Civil Rights Act, 1965 Department of Housing and Urban Development Act, 1966 Demonstration Cities and Metropolitan Development Act, 1966 Department of Transportation Act, 1968 Intergovernmental Cooperation Act, 1968 Civil Rights Act, 1969 Environmental Policy Act, and the 1974 Housing and Community Development Act.

3. Additional resources included historical photographs and maps, visual and questionnaire surveys, and numerous interviews, both informal and by formal deposition. Examples from the most recent studies, in Norfolk, Alexandria, and Kansas City, are more frequently cited here.

4. The places where discriminatory disparities were found in the provision of municipal facilities to black residential areas included Lowndes County and Selma, Alabama; Pine Bluff, Arkansas; Sanford, Florida; Ocilla, Georgia; Shreveport, Louisiana, Baltimore County, Maryland; Drew, Gulfport, Itta Bena, Senatobia. Shaw and West Point, Mississippi; Las Vegas, Nevada; Mt. Laurel, New Jersey; Jackson, Tennessee; Fairfax County and Greenville County, Virginia.

REFERENCES

Bowden, D. L., & Palmer, J. L. (1984). Social policy: Challenging the welfare state. In J. L. Palme & I. V. Sawhill (Eds.), *The Reagan record: An assessment of America's changing domestic priorities* (pp. 177-215). Cambridge: Ballinter.

Cusick, P. A., Gerbing, D. W., & Rossel, E. L. (1979). The effects of school desegregation and other factors on white flight from an urban area. *Educational Administration Quarterly, 15*(2), 35-49.

Farley, R., Schuman, H., Biachi, S., Colosanto, D., & Hatchett, S. (1983). Chocolate city, vanilla suburbs: Will the trend toward racially separated communities continue? In M. Baldassare (Ed.), *Cities and urban living* (pp. 292-315). New York: Columbia University Press.

Farley, R., & Tauber, K. E. (1968). Population trends and residential segregation since 1960. *Science, 159*(3818), 952-956.

Feagin, J. R., & Feagin, C. B. (1978). *Discrimination American style: Institutional racism and sexism.* Englewood Cliffs, NJ: Prentice-Hall.

Foley, D. L. (1973). Institutional and contextual factors affecting the housing choices of minority residents. In A. H. Hawley & V. P. Rock (Eds.), *Segregation in residential areas* (pp. 85-147). Washington, DC: National Academy of Sciences.

Hammer, Siler, George. (1976). *Final NEA study report to the city of Alexandria, Virginia.* Washington, DC: Government Printing Office.

Jones, E. R. (1982). *The differential impact of income and preference on residential segregation* (Planning paper 82-002). Urbana-Champaign: University of Illinois Bureau of Urban and Regional Planning.

Judge finds Yonkers segregates schools. (1985, November 21). *New York Times*, p. 1.

Kain, J. F. (1974). Housing segregation, black employment and metropolitan decentralization: A retrospective view. In G. M. von Furstenberg, B. Harrison, & A. R. Horowitz (Eds.), *Patterns of racial discrimination: Vol. 1 Housing* (pp. 5-18). Lexington: Heath.

LeGates, R., & Hartmann, C. (1982). Chapter 14. In *Displacement: How to fight it.* Berkeley, CA: National Housing Law Project.

Long, L., & DeAre, D. (1981, September). The suburbanization of blacks. *American Demographics,* pp. 17-21, 44.

National Low Income Housing Information Service. (1986). *Displacement Forum, 1*(1), 2.

Nelson, K. P. (1979). *Recent suburbanization of blacks: How much, who, and where.* Washington, DC: HUD.

Orfield, G. et al. (1980). A social science statement: School segregation and residential segregation. In W. G. Stephen & J. R. Feagin (Eds.), *School desegregation past, present, and future* (pp. 231-247). New York: Plenum.

Palmer, J. L., & Sawhill, I. V. (Eds.). (1984). Overview. In J. L. Palmer & I. V. Sawhill (Eds.), *The Reagan record: An assessment of America's changing domestic practices* (pp. 1-30, 206). Cambridge: Ballinger.

Pearce, D. M. (1981). Deciphering the dynamics of segregation: The role of schools in the housing choice process. *Urban Review, 13*(2), 85-102.

Pettigrew, T. F. (1973). Attitudes on race and housing: A social-psychological view. In A. H. Hawley & V. P. Rock (Eds.), *Segregation in residential areas* (pp. 21-84). Washington, DC: National Academy of Services.

Rabin, Y. (1970, August). *The effects of development control on housing opportunities for black households in Baltimore County, Maryland* (Report to the U.S. Commission on Civil rights).

Rabin, Y. (1975). *Housing segregation in Philadelphia and the Whitman Park Housing Project: The role of city and federal actions* (Report to Community Legal Services of Philadelphia).

Rabin, Y. (1980). Federal urban transportation policy and the highway planning process in metropolitan areas. *Annals of the American Academy of Political and Social Science, 451*, 21-35.

Rabin, Y. (1983). The final question: Who benefits? *Planning, 49*(10).

Rabin, Y. (1984). Suburban racial segregation and the segregative actions of government: Two aspects of metropolitan population distribution. In *A sheltered crisis: The state of fair housing in the eighties* (pp. 31-53). Washington, DC: G.O.P.

Streitweiser, M.L., & Goodman, J. L., Jr. (1983). A survey of recent research on race and residential location. In *Population Research and Policy Review, 2* (pp. 253-283). Amsterdam: Elsavier.

Tauber, K. E. (1984). *Racial residential segregation, 28 cities, 1970-1980* (CDE Working Paper 87-12). Madison: University of Wisconsin.

Tauber, K. E., & Tauber, A. E. (1965). *Negroes in cities.* Chicago: Aldine.

Yinger, J. (1979). Prejudice and discrimination in urban housing. In P. Mieszkowski & M. Straszheim (Eds.), *Current issues in urban economics* (pp. 430-468). Baltimore: Johns Hopkins.

10

The Implementation of the Federal Mandate for Fair Housing

BETH J. LIEF
SUSAN GOERING

RESIDENTIAL SEGREGATION by race is neither a natural nor an inevitable phenomenon. Although prejudice on the part of individual sellers, buyers, and renters undeniably was and is a factor in segregated housing patterns, such patterns are primarily the product of two other sets of forces. Federal, state, and local government programs, on the one hand, interacted with and reinforced the segregative and discriminatory practices of a variety of private institutions involved in the housing market, such as real estate interests and financial institutions, on the other. Yet perhaps "because of the extensive nature of its involvement in housing and community development, the Federal Government . . . has . . . been most influential in creating and maintainig urban residential segregation" (U.S. Commission on Civil Rights, 1975; Weaver, 1983).

This chapter will focus on federal laws and regulations that attempted to ameliorate segregation in housing and their implementation on racial segregation. It will focus on the United States Department of Housing and Urban Development (HUD) and its predecessor agencies[1] covering a period from 1962 to the present. An analysis of the policies of the last quarter century would be incomplete without an initial discussion of the federal government's active promotion of a dual housing market[2] from the 1930s well into the 1950s. This chapter will then describe several of the federal government's initiatives through legislation and executive order to dismantle the dual housing market during the 1960s, and will examine the impact of those initiatives on several housing programs administered by HUD. Finally, the chapter will analyze some of the major deficiencies in HUD's administration of these programs and its enforcement of the fair housing laws.

The history of the federal government's role in promoting equal access to integrated housing demonstrates that a successful effort to eliminate the complex and entrenched forces that perpetuate the dual, segregated housing market—even one launched on a federal scale—requires strong, affirmative policies to undo decades of practice and a firm commitment of resources toward enforcement of those policies. Federal initiatives have fallen short on both counts.

EARLY FEDERAL GOVERNMENT POLICIES THAT CAUSED RACIAL SEGREGATION

The federal government's role[3] in creating a dual housing market dates back at least to the early decades of this century.[4] The collapse of the private housing market during the depression brought massive government involvement in the housing industry. During the New Deal, in order to salvage the home building and financing industries, the new federal agencies that were created for this purpose adopted and reinforced the racial practices of the private real estate and mortgage lending institutions (Orfield, 1974).

Prior to and continuing after the 1930s, individual white home owners and neighborhood associations in large sections of the country entered into covenants prohibiting the sale of housing to blacks and others groups that were perceived by them to be "undesirable." A number of states enforced such covenants in their courts. However, the explicit policies and practices of the Federal Housing Administration (FHA), from its inception in 1934[5] until at least the early 1950s, not only encouraged the use of these racial restrictive covenants but also, in light of the lack of national appraisal standards in those early years, helped institutionalize the concept of racial homogeneity as a critical indicia in determining and maintaining property values.

During the 1930s, the FHA actually drafted a model restrictive covenant for use in home sales that it financed, and strongly encouraged its use to maintain the racial homogeneity of neighborhoods.[6] FHA's underwriting of homes whose deeds contained racially restrictive covenants continued unmodified for two years after the United States Supreme Court held the enforcement of such covenants unconstitutional in the 1948 case of *Shelley v. Kraemer*.[7] Even then, the FHA only stopped insuring mortgages on property encumbered by covenants entered into after February 15, 1950 (United States Department of Housing and Urban Development [HUD], 1973).

The FHA's *Underwriting Manuals* referred to "the infiltration of inharmonious racial or nationality groups" as "adverse" to neighborhood stability[8] and advised appraisers to lower the rating of properties in mixed neighborhoods, often to the point of rejection.[9] Although the FHA removed explicitly racist language from its manuals in the 1950s, later manuals continued to accord value to "compatibility among neighborhoods' occupants" and warned of the risk of "dissimilar" user groups.[10]

The FHA also redlined black neighborhoods. Based on its appraisal standards, the FHA "generally withheld insurance from existing housing in central city areas" because those areas "occupied largely by minority groups had an unfavorable economic future" (Romney, 1970). The effect of these policies was to finance white flight from the inner city, to deny blacks a similar opportunity to become suburban home owners, and to ensure a rapid increase in segregation along urban/suburban lines. FHA policies meant that blacks were denied mortgages for the only housing available to them in inner city black neighborhoods, which caused deterioration in the black housing stock and precluded blacks from accumulating the wealth that accrues from home ownership.[11]

The impact of these policies was enormous in both the public and private housing industries. From the 1930s until 1959, the FHA financed three out of every five homes purchased and enabled millions of families to purchase homes. "FHA mortgage insurance revolutionized home financing by guaranteeing payment on mortgages which met the agency's standards of housing quality and appraised market value. With the risk eliminated, [private] lenders were willing to accept lower interest rates and much longer periods of repayment" (Orfield, 1974). But virtually all of these FHA subsidized homes were in the suburbs and virtually all housed whites.[12] Less than 2% of FHA insured loans were made to blacks from the mid-1940s through the end of the 1950s (U.S. Commission on Civil Rights, 1973, 1975).

The FHA programs were important beyond the absolute number of houses they financed. FHA publications touted the agency's practices as trend-setting models for private realtors, lenders, and appraisers (FHA, 1960a, 1960b). Those practices created a national set of standards in the home mortgage finance and sales industry. Indeed, FHA publications in the 1930s that encouraged racial segregation were still cited in the training manuals of the major private appraisal associations as late as the 1970s.[13]

The explosive growth of these virtually all-white suburbs during the 1950s and 1960s was also fueled by federally subsidized highways. The highways opened vast tracts of land along the urban periphery for development, making it easier for large segments of the population that worked in the city to escape to the all-white suburbs.

Moreover, the placement of the highways themselves frequently resulted in segregated housing patterns. All too often, highway rights-of-way isolated or eliminated black residences in white areas, or were otherwise placed in a manner that contributed to increased concentrations of blacks in the inner city. New roads frequently uprooted suburban minority communities, forcing minority suburbanites to relocate in the central city (Grier & Grier, 1977; U.S. Commission on Civil Rights, 1974).[14]

The funding and administration of the low-income public housing program after its inception in the 1930s[15] similarly played a substantial role in segregating the vast majority of blacks within the inner city. Semiautonomous local public housing authorities ("PHAs"), with the federal government's funding and approval, regularly segregated housing projects by races, creating "white projects" and "black projects" (Falk & Franklin, 1976).[16]

Two separate waiting lists for occupancy were typically employed to perpetuate the system. That practice continued in many locations well into the 1960s and even 1970s,[17] long after 1954 when the U.S. Supreme Court held in the school desegregation cases that legally compelled or sanctioned segregation by state and local agencies was unconstitutional.

Moreover, the site selection of the projects themselves promoted segregation within the cities. Housing projects whose occupancies were expected to be largely black were consistently built within the inner city in areas of high minority concentration, often as part of a slum clearance or urban renewal project (Falk & Franklin, 1978). Regulations that governed urban renewal programs encouraged the building of replacement housing in the same neighborhood and accorded priority for such housing to the former residents of the neighborhood. Adoption of this policy, while perhaps a justifiable attempt to ensure housing for those displaced by urban renewal, meant forgoing a real opportunity to integrate urban and suburban neighborhoods.[18] "In virtually all metropolitan areas, the location of public housing accentuated the concentration of minority groups in central cities" (Grier & Grier, 1966; U.S. Commission on Civil Rights, 1975).[19] In some cities, "the policies pursued by [the public housing authorities] with the [federal] Government's blessing, actually created segregated residential patterns and concentrations of minority poor where they had not existed before" (U.S. Civil Rights Commission, 1975). By 1962, close to 80% of all federally subsidized public housing projects were occupied by only one race (Grier & Grier, 1966).

Meanwhile, the federally funded Urban Renewal Program, established by the Housing Act of 1949 to facilitate urban renewal and revitalization in local communities, operated during the decades before 1960 to perpetuate and increase racial segregation. Where there were pockets of black residents interspersed in predominantly white neighborhoods, too often "the definition of successful renewal became efficient removal of blacks from an area and their speedy replacement by higher-income whites or businesses" (Orfield, 1974). More egregious still was the regular pattern of relocating blacks into areas of higher black concentration than those areas from which they were displaced. The local agencies charged with relocating residents uprooted by urban renewal frequently maintained separate lists of replacement housing for white and black displacees, thus perpetuating and reinforcing preexisting segregation in the housing market.

By 1960, the dual housing market was firmly entrenched, in large part as a result of federal government programs that financed the development of neighborhoods along racial lines and policies that institutionalized in both the private and public sectors a dual housing market with separate areas, separate rules, and separate institutions serving whites and blacks. That dual housing market possessed a momentum and self-reinforcing character that would make it seemingly impenetrable even in the face of the best of federal government intentions.

THE SHIFT TOWARD FAIR HOUSING: THE FEDERAL MANDATE

Beginning in the 1960s, the federal government evidenced a shift in policy and initiated a series of attempts to mandate fair housing. Executive and legislative initiatives, of which only the major ones are discussed here, took various forms and, during the late 1960s and early 1970s, became increasingly aggressive but they failed, individually and as a group, to achieve a significant impact in dismantling the dual housing market, creating equal access to housing, and achieving housing integration.[20]

EXECUTIVE ORDER 11063 (1962)

On November 20, 1962, President John Kennedy issued Executive Order 11063, which prohibited discrimination in all housing that received federal aid after that date. It represented the first explicit official statement of national policy that the federal government was opposed to discrimination in housing.

The Executive Order applies to property (1) owned or operated by the federal government (mainly housing on military or government installations and units repossessed by the FHA and the Veterans' Administration (VA) and (2) that receives some form of government assistance, either through loans or other contributions (e.g., public housing, urban renewal). The order also prohibits discrimination by lending institutions, but only as to loans insured or guaranteed by the federal government. Violators are subject to a number of penalties, including cancellation of contracts or exclusion from other governmental assistance. The scope of the order, which covers less than 1% of the nation's housing, is extremely narrow (Comment, 1969), since by 1962 most of the nation's housing was financed through loans made by private lending institutions, that is, not ones guaranteed by the FHA and the Veteran's Administration (U.S. Commission on Civil Rights, 1973). Not only was the Order's scope narrow, enforcement was minimal. As of 1966, responsibility for enforcement of the Executive Order rested chiefly with previously existing housing agencies, aided by only a small President's Committee on Equal Housing Opportunity, which, despite its national responsibilities, had less paid staff than was available to enforce antidiscrimination housing laws in the state of New York (Grier & Grier, 1966).

In 1980, HUD issued regulations[21] to carry out the requirements of Executive Order 11063 as it pertained to properties and facilities owned or operated by the federal government, assisted by HUD funding, or encumbered by loans insured by HUD.

Under the regulations "discriminatory practices" are broadly defined essentially to include any practices that have the effect of denying equal housing opportunities or that substantially impair equal access to benefits of HUD's housing programs. Noncompliance with HUD's affirmative fair housing marketing requirements or a formal finding of a violation of the Fair Housing Act of 1968 (Title VIII) also constitute a violation of the Order.[22]

The regulations provide for an administrative complaint[23] and disposition process,[24] and for compliance reviews by HUD's Regional Offices of Fair Housing and Equal Opportunity (FHEO),[25] to determine whether the respondent is in compliance with E.O. 11063.

Sanctions and penalties that may be imposed include such measures as cancellation of the contract, refusal to approve a lender, a determination of ineligibility, and suspension from further participation in HUD programs.[26] In appropriate cases, HUD may refer the case to the U.S. Attorney General for appropriate civil or criminal action.[27]

TITLE VI OF THE CIVIL RIGHTS ACT OF 1964

The first major civil rights legislation to affect housing was Title VI of the Civil Rights Act of 1964, which prohibits discrimination in any program or activity receiving federal financial assistance, including housing discrimination in federally assisted programs such as public housing and urban renewal.[28] For example, when local governments use federal financial assistance to operate low-income housing, they are prohibited by Title VI from discriminating on the basis of race, color, or national origin in renting existing housing or choosing sites for new housing. Despite its wide reach, Title VI contains a major caveat. Programs of insurance and guaranty are outside its parameters,[29] and loan guaranty programs such as the FHA, VA, and the Farm Mortgage Home Administration (FMHA) mortgage programs, which cover millions of housing units, are therefore beyond its strictures.

Responsibility for enforcing Title VI rests with the various federal agencies that provide assistance to recipients. Title VI provides that compliance may be effected by the termination of assistance or by "any other means authorized by law,"[30] including the initiation of litigation by the Justice Department.[31]

Short of judicial intervention, administrative mechanisms are available under Title VI. Federal agencies were required to and have promulgated Title VI regulations that provide steps to gain voluntary compliance by negotiations and conciliation between the parties and agreements to avoid future discrimination. Where voluntary compliance efforts fail the agencies may act administratively temporarily to withhold or even to terminate federal aid.

The Housing and Home Finance Agency (HHFA), the predecessor to HUD, issued the first regulations related to housing programs under Title VI.[32] They contained a relatively broad proscription against discrimination in federally assisted housing programs.[33] The regulations' proscription against discrimination included a prohibition against methods of administration that had the direct or indirect effect of limiting access to or segregating persons on the basis of race. Thus the regulations covered, for example, the "freedom of choice" plans used by many local housing authorities for assigning tenants to public housing, as those had the effect of allowing applicants to choose to live in public housing where the race of the other residents was the same as their own.

The 1965 Title VI regulations required that every application and contract for financial assistance to carry out federally assisted programs contain assurances that the programs or activities would be administered in compliance with Title VI.[34]

The regulations provided several tools by which the agency officials with principal responsibility for administering housing assistance programs could ensure compliance with Title VI. First, each recipient of federal housing funds was required to submit compliance reports with sufficient regularity for the agency official to check continuing compliance.[35] Second, HUD officials were required "from time to time" to review the practices of recipients to determine whether they were complying.[36] Third, any persons or class of persons who believed themselves to be victims of discrimination could file a written complaint within 90 days from the date of discrimination.[37] Current regulations allow 180 days to file a complaint.[38]

Although the initial regulations had no reference to affirmative efforts to eliminate segregation or discrimination, by 1973, HUD in its Title VI regulations required affirmative action to overcome the effects of prior discrimination under two circumstances—(1) where the recipient had previously discriminated[39] and (2) where previous discriminatory practice or usage had lingering discriminatory effects.[40] HUD also required that instruments transferring real property acquired through federal assistance contain a covenant assuring nondiscrimination.[41]

In practice, the reach of these regulations has been limited. For example, "in fiscal year 1977, 21 percent of all HUD's compliance reviews of recipients focused on private sponsors and owners, representing less than one percent of these participants in HUD programs. Fifty-six percent of HUD's compliance reviews focused on local public housing authorities, representing only three percent of these participants" (U.S. Commission on Civil Rights, 1979).

TITLE VIII OF THE
CIVIL RIGHTS ACT OF 1968
(The "Fair Housing Act")

The only federal statute devoted exclusively to the elimination of discrimination in housing is Title VIII of the Civil Rights Act of 1968,[42] also referred to as the "Fair Housing Act," which prohibits discrimination based on race, color, religion, sex, and national origin in the sale or rental of housing.[43] Unlike Title VI, it covers activities of private as well as governmental segments of the real estate industry, including real estate brokers, builders, apartment owners, sellers, and mortgage lenders, and federally owned and operated dwellings and dwellings constructed, rehabilitated, or purchased with federally insured loans and grants.

Title VIII prohibits a wide variety of discriminatory activities on the basis of race, color, religion, sex, or national origin. These prohibited activities include the refusal to sell or rent a dwelling, and[44] discrimination in the terms, conditions, or privileges of the sale or rental of a dwelling,[45] including negotiations, requirements for down payments, and credit checks.

Also prohibited are any:

—indications of preference, limitation, or discrimination in advertising[46]
—representation to a person or persons that a dwelling is unavailable[47]
—denial of a loan for purchasing, constructing, improving, or repairing a dwelling[48]
—discrimination in setting the amount or other conditions of a real estate loan[49]
—denial of access to or membership in any multiple-listing service or real estate brokers' organization[50]

Title VIII also prohibits such forms of discrimination as "blockbusting"—convincing owners to sell property on the grounds that minorities are about to move into a neighborhood—and "steering"—the process of directing a racial, ethnic, or religious group into a neighborhood in which members of the same group already live.[51]

The Act also provides that it is unlawful for any bank, building and loan association, or other institution engaged in making real estate loans to deny a loan or other financial assistance for purchasing, constructing, repairing, or maintaining a dwelling or to discriminate against borrowers in fixing the amount, interest rate, duration, or other terms or conditions of such a loan because of an applicant's race, color, religion, national origin, or sex.[52]

Responsibility for overall administration of Title VIII rests with the Department of Housing and Urban Development, which has authority to investigate and conciliate complaints of housing discrimination.[53] Title VIII imposed a legal obligation on the executive branch to promote equal opportunity affirmatively. Section 808 (e) (5) requires that the secretary of HUD "administer its programs and activities relating to housing and urban development in a manner affirmatively to further the policies" of fair housing.[54]

The Department of Justice is the only unit of the executive branch to which Congress has assigned enforcement authority under Title VIII.[55] Section 813 gives the Attorney General the authority to litigate when there is a pattern or practice of housing discrimination or where issues of housing discrimination are of general public importance. Section 812 provides for enforcement by private persons in federal or state courts.[56] Thus complainants may also proceed to federal court themselves to enforce their rights, simultaneously or independently of pursuing their administrative remedies.[57] Originally the Fair Housing Act contained a provision requiring courts to expedite the judicial proceedings by the Department of Justice and private citizens in every way,[58] but that provision was repealed in 1984.[59]

In January 1969, HUD issued regulations pursuant to Title VIII[60] for the purpose of carrying out its responsibility with respect to any complaint filed with the Department under Sections 804, 805, and 806 of Title VIII.[61] The scope of these regulations was narrow and basically procedural, placing HUD in the position of reacting to complaints filed by private parties rather than moving affirmatively to dismantle residential segregation. Provision was made for HUD to refer any complaint to a state or local agency if that agency provides rights and remedies substantially equivalent to those provided under Title VIII.[62]

If HUD decides to attempt resolution of the complaint, it may do so only by informal methods of conference, conciliation, and persuasion, which may proceed simultaneously with further investigation.[63] HUD may not impose civil penalties for noncompliance with Title VIII. In other words, HUD has no enforcement authority. In the event a settlement is reached, HUD retains authority to review compliance with any settlement agreement at a later date, but lacks power to enforce compliance.[64]

If attempts at obtaining voluntary compliance are terminated in a case where there is probable cause to believe the respondent had committed a discriminatory housing practice, HUD may (1) recommend that the Attorney General seek relief under Title VIII against a pattern or practice where a group of persons raises an issue of general public importance, (2) refer the matter to the Attorney General for such other action as he or she may deem appropriate, (3) institute proceedings under E.O. 11063 or Title VI where appropriate, or (4) contact other federal agencies having any interest in the proceedings.

One major weakness in HUD's effort to combat residential discrimination is that its efforts under Title VIII are focused largely on processing complaints rather than on initiating compliance reviews. Conducted systematically, compliance reviews would have the potential for greater impact on systemic discriminatory practices than does the individual complaint investigation and conciliation process.[65]

Beginning in 1971, three years after Title VIII became law, HUD began to issue a series of regulations that authorized more affirmative actions to implement the mandate of Title VIII, some of which are discussed immediately below.

Affirmative Fair Housing Marketing Regulations

In 1972, HUD issued regulations requiring affirmative fair housing marketing under HUD subsidized and unsubsidized housing programs.[66] For example, regulations set forth as HUD's policy the administration of its FHA[67] programs affirmatively to achieve a condition in which individuals of similar income levels in the same housing market area have a like range of housing choices available to them regardless of race, color, religion, sex, or national origin. Toward that end, applicants for participation in all HUD-assisted housing programs are required to use affirmative fair housing marketing plans (AFHM plans) in soliciting buyers and tenants, in determining their eligibility, and in conducting sales and renewal transactions.[68]

According to the regulations, such AFHM plans must be designed to

(a) Carry out an affirmative program to attract buyers or tenants, regardless of sex, of all minority and majority groups to the housing for initial sale or rental by publicizing to minority persons the availability of housing opportunities.
(b) Maintain a nondiscriminatory hiring policy for staff engaged in the sale or rental of properties.

(c) Instruct all employees and agents in writing and orally in the policy of nondiscrimination and fair housing.
(d) Specifically solicit eligible buyers or tenants reported to the applicant by the Area or Insuring Office.
(e) Prominently display the Department-approved Fair Housing Poster and include in any printed material the Department-approved Equal Housing Opportunity logo or slogan or statement.
(f) Post in a conspicuous position on all FHA project sites the Equal Housing Opportunity logo or slogan or statement.[69]

In addition, an important requirement is for developers and sponsors to state in their plans the anticipated results of the plans in terms of the number or percentage of dwelling units they will sell or rent to minorities.[70]

A major weakness of these AFHM regulations is that they do not apply to existing FHA-insured or subsidized projects, even though racial data collected on existing subsidized, multifamily units show extensive segregation. Moreover, the AFHM plan requirements apply only to initial occupancy of the housing covered by these regulations, thus limiting the potential effect of the regulations long term. Finally, the regulations cover only those particular projects and subdivisions that a builder develops under FHA programs and that same builder is not required to market all its housing affirmatively (HUD, 1976; U.S. Commission on Civil Rights, 1973, 1979). (A 1976 HUD-sponsored study also summarized the inadequacies of the AFHM program; HUD, 1976, pp. 9-13.)

In 1979, as part of its mandate under E. O. 11063 and Title VIII HUD promulgated regulations[71] to establish procedures for determining whether the developers of HUD subsidized housing are in compliance with their HUD-approved Affirmative Fair Housing Marketing (AFHM) Plans.[72] The Fair Housing and Equal Opportunity (FHEO) Division of HUD's Area Office is responsible for monitoring the AFHM and providing technical assistance during the development and implementation of the plans.[73] HUD's Regional FHEO is responsible for determining whether the developers are in compliance with their plans and AFHM regulations and for ordering postmarketing changes.[74] That office conducts compliance reviews to examine such matters as the developer's sales and rental practices, programs to attract minority and majority buyers and renters regardless of sex, and data regarding the size and location of units, the race and sex of buyers and renters, and the race and sex of staff engaged in sales or rentals. (24 CFR Sec. 108.40(c).)

Fair Housing Assistance Program

As mentioned above, Title VIII provides, in effect, that HUD may refer a complaint of housing discrimination to a local agency for investigation and resolution wherever a state or local fair housing law provides rights and remedies for discriminatory housing practices that are substantially equivalent to those provided in Title VIII.

In 1980, HUD implemented a program to assist the local bodies in handling the complaints referred to them by HUD. The Fair Housing Assistance Program[75] provided assistance to these state and local agencies, including assistance in complaint processing, training and technical assistance, data support systems, and funding for innovative projects. The intent of the programs is to build a coordinated intergovernmental enforcement effort to further fair housing and provide incentives for involvement by local agencies.[76]

Fair Housing Advertising

HUD's regulations on Fair Housing Advertising[77] were promulgated in 1980 on the authority of Title VIII, as amended.[78] These regulations describe the criteria that the Assistant Secretary for FHEO uses in evaluating compliance with Title VIII during investigations of complaints alleging discriminatory housing advertisements.[79] The implementation of certain fair housing policies and practices, for example, use of the Equal Housing Opportunity logo, is evidence of compliance with Title VIII.[80]

OTHER FEDERAL FAIR HOUSING LAWS

In addition to Title VI and Title VIII, Congress enacted two statutes specifically targeted to bolstering nondiscrimination in the mortgage and credit industry—the Equal Credit Opportunity Act of 1974[81] and the Community Reinvestment Act.[82] Although not the responsibility of HUD to administer or enforce,[83] these statutes prohibit racial discrimination in credit and encourage banks to issue credit to low-income and moderate-income communities, sites that historically have found mortgage and credit availability lacking.

THE MISSED OPPORTUNITY: WHAT WENT WRONG

Despite the apparently clear executive and legislative mandates that racial discrimination in housing be ended, segregation has in fact increased in many cities since 1962. For instance, in the decade between 1970 and 1980, in the 28 cities with a population over 100,000, one study found that integration on a block-by-block basis actually declined on average (Taeuber, 1983).

This is particularly surprising because one would expect that the substantial impact on housing availability and location occasioned by federal programs since 1962, coupled with the federal mandates prohibiting housing discrimination, would have effected some positive change in residential segregation.

During the 1960s and until 1972, billions of dollars were administered by HUD in projects that directly affected the housing market. During that time, hundreds of thousands of housing units were made available as a result of federal financial assistance. For example, between 1968 and 1972 alone, two HUD subsidized programs—the FHA section 235 program for low-income home owners and the FHA Section 236 program for low-income renters made over 600,000 units available. In 1970, subsidized housing starts generally constituted 29.3% of that year's total starts (Weaver, 1983).

The reasons for the paradox are complex. This section examines the shortfalls of the Department of Housing and Urban Development, whose potential for decreasing segregation was the greatest.

The Department of Housing and Urban Development is the major federal agency responsible for improving housing conditions in this country. It does so by providing assistance to citizens, developers, public housing authorities, and private nonprofit housing agencies for the financing and production of new housing, preservation of available housing, leasing of housing, and improvement of substandard housing. In addition to the lower-income rental assistance program, the low-income public housing program, the mortgage interest subsidy program, and mortgage insurance programs, which will be the focus of the discussion below, HUD's programs also include the community development block grant program, the urban development action grant program, and the comprehensive planning assistance program, all of which provide opportunities for promoting fair housing.

Several common themes emerge from an analysis of HUD's administration of its diverse federal housing programs during the 1960s and 1970s as to why the impact on the dual housing market was so surprisingly limited.

THE INHERENT LIMITATIONS IN HUD'S AUTHORITY

The legislative mandates for fair housing discussed in the previous section had inherent limitations. For example, despite its broad coverage, Title VIII limits HUD's authority. Even when compliance reviews for individual complaints demonstrate evidence of discrimination, HUD is limited to a conciliation role and cannot hold administrative hearings or issue orders to have offenders cease illegal conduct or undertake other affirmative action to eradicate the effects of past discriminatory conduct. The only recourse HUD has is to refer a complaint to the Department of Justice for prosecution in federal court.[84] The only other recourse for the private complainant is to go to court—an expensive and often time consuming process, particularly when seeking classwide or broad relief. Moreover, already overburdened courts are saddled with an additional burden in an area, housing, in which they possess little experience or expertise.

These shortcomings in Title VIII have tangible effects. By 1974, HUD regional offices had been receiving well over 2,000 complaints under Title VIII each year, but its processing of them had been found to be ineffective (U.S. Commission on Civil Rights, 1974). The U.S. Commission on Civil Rights studied 1,601 complaints handled by HUD during a nine month period in 1972 and 1973. Only slightly over a fifth of the complaints went to conciliation and just over one-half of those were conciliated successfully. In contrast, of the cases that HUD referred to state and local agencies, approximately one-fifth went to conciliation with nearly 96% of those being successfully conciliated (U.S. Commission on Civil Rights, 1974). Approximately four-fifths of the sample, or 1,339, were dropped without any relief to the complaining party. The shortcomings are, at least in part, attributable to an understaffing and to a cumbersome conciliation process that lacks enforcement teeth.

The failure to grant HUD rule-making authority to prescribe objective guidelines for nondiscriminatory activity in the housing industry is another defect. These could have established criteria to determine what conduct is *prima facia* discriminatory, unless properly validated by business necessity, comparable to criteria for testing for employee selection in the equal employment context. Such regulations, if given the force of law in litigation, could significantly have influenced private actions (Falk & Franklin, 1976).

Since Title VIII first passed, several efforts to strengthen it have failed in Congress. For instance, a legislative amendment offered in 1980, over 10 years after the passage of Title VIII, to confer on HUD the authority to hold administrative hearings and issue orders, reviewable only by a federal court, failed to pass both houses. The principal thrust of Senate Bill 506, the Fair Housing Amendments Act of 1980, was to strengthen the enforcement mechanism of Title VIII by *inter alia* providing for a three person fair housing commission and hearings by an administrative law judge who could order relief that included compensatory damages and civil penalties up to 10,000 dollars.[85]

To HUD's credit, its secretaries under the Johnson, Nixon, Ford, and Carter administrations have uniformly criticized the existing Fair Housing Act for its inadequate administrative enforcement mechanism. Provisions for such mechanisms would have lent credibility to HUD's enforcement efforts and shifted the burden of time and cost now imposed on the private complainant to the government, which has greater resources.

But HUD's failure to attack fully the problems of racial segregation in housing full force is due not only to real legal limitations on its authority, but also to its failure to take an expansive view of what authority it had. For example, although Executive Order 11063 barred discrimination only in federally financed housing, that is, only 1% of the nation's housing, the FHA took a more restrictive view of the Executive Order's scope than was necessary and declined to apply it either to the single-family loans that were issued prior to that date, or to prospective single-family loans unless insurance was approved in blocks of four homes or more. As will be discussed below, what authority was available to HUD was too often neither interpreted nor applied with the force necessary to effect housing patterns on a substantial scale.

HUD'S UNWILLINGNESS TO ACT TO CHANGE THE OPERATION OF PRIVATE MARKET FORCES AND ITS WILLINGNESS TO RELY ON RACE-NEUTRAL PLANS OR PROGRAMS

The development of largely black inner cities and white suburbs was the product of several interrelated public and private decisions. "One of the most central was the public decision to deal with the housing shortage chiefly through the private enterprise system. Such government mechanisms as were mobilized to aid in the task, especially the mortgage guaranty provisions of the FHA and VA, were all directed to encourage the efforts of private enterprise" (Grier & Grier, 1966).

The powerful institutional forces in the private sector that determine where blacks reside and what form metropolitan growth takes have been well documented. These forces include real estate agents who have controlled the market by steering, controlling the access of black brokers to white listings, and blockbusting; financial institutions that have redlined and otherwise discriminated in granting loans and undervaluing houses in appraisals in black neighborhoods; and builders and the construction industry who discriminate either actively or passively by allowing community forces to determine the racial occupancy of their projects (U.S. Commission on Civil Rights, 1974). Race-neutral or "race blind" practices of HUD that continued to finance these private actors served only to perpetuate, if not strengthen, the dual, segregated housing patterns in the country.[86]

In the 1960s and 1970s, FHA's own appraisal practices, linked to those of private real estate institutions, effectively redlined inner-city areas that were predominantly black. In 1967 the FHA itself studied its mortgage insurance practices for minorities across the country and found it was insuring virtually no loans to minorities in most housing markets.[87]

Through at least the mid-1960s or, in some cities the mid-1970s, for instance, the FHA's traditional Section 203 mortgage guarantees were almost impossible to obtain by a black in a black area; Section 203 housing was virtually unattainable for blacks in white areas.[88] Despite those findings, the FHA made no significant effort to promote the availability of mortgages for homes in white areas to blacks in order to change the segregationist real estate practices in the private market that was relying on FHA financing.[89]

HUD's acquiescence in the operation of private market forces is perhaps best characterized by its implementation of its Section 235 program.[90] The new housing in the Section 235 program was located primarily in suburban areas and was bought by white families, while most of the existing housing purchased under the 235 program was in predominantly inner-city black areas or changing neighborhoods and was purchased mostly by black families. As it turned out, the existing housing located in the central city was often substandard housing, and the subject of graft in the 235 program. Thus black families suffered disproportionately from the abuses that occurred under that program (U.S. Commission on Civil Rights, 1971). Analyzing the anatomy of segregation in 235 housing, the U.S. Commission on Civil Rights in June 1971 asked,

> Why has the traditional pattern found in the housing market in general been repeated in the 235 program? A strong arsenal of civil rights law exist to protect minority home seekers from discrimination in the 235 program as well as in all other housing. Further, the economic rationale for the dual housing market that exists generally has no application here. All eligible families, minority or majority, are required, by statute, to be in the same income range, and all housing, whether suburban or inner-city, whether new or existing, is required, again by statute, to be within the same cost limitations. Nevertheless, the dual housing market persists in the 235 program—a market that is separate and unequal.

The Commission concluded that "the answer [lay] in the way in which the program has been administered. Each of the actors and institutions involved in the 235 process—real estate brokers, builders, mortgage lenders, interested community groups, Government, and the buyer himself—had played a role in producing the segregated, unequal product" (U.S. Commission on Civil Rights, 1971).

Discussing the role of FHA, the Report concluded that:

> FHA officials, moreover, even though aware of the segregated housing pattern that has developed under the 235 program, [had] failed to take even minimal steps to change it, despite their legal obligation to do so. Not until 1971, did HUD begin to collect racial data on Sec. 235 buyers. FHA and HUD's Office of Equal Opportunity [relied] mainly on the process of complaints as the mechanism for discovering and eliminating discriminatory practices. The central office in Washington . . . failed to provide local FHA offices with instructions for affirmative action aimed at broadening the range of housing choice for minority families. Local FHA officials [were] reluctant to take such action, in some cases, for fear that the central office would not back them up.
>
> Thus FHA, the agency charged by Congress with responsibility for administering the 235 program, . . . abdicated its responsibility and, in effect, . . . delegated it to members of the private housing and home finance industry. In view of the traditional policies and attitudes that . . . predominated in [the real estate] industry, the pattern of separate and unequal housing under the 235 program [had] been inevitable. (U.S. Commission on Civil Rights, 1971)

The 1971 Commission warned that "until FHA abandons its current passive role and becomes a vigorous champion of the rights of minorities and of lower-income families generally, this pattern is unlikely to change" (U.S. Commission on Civil Rights, 1971). Unfortunately, the impact of these policies was dramatic because the 235 program operated on a large scale from 1969 through 1972.

HUD's approach to tenant placement in the low-income public housing projects it funded is another example of HUD's passivity, which permeated all its housing programs. The Public Housing Administration (PHA), the federal agency primarily responsible for administration of public housing programs prior to 1962, allowed local authorities to use a "free choice" plan to comply with Title VI. Under that plan, all applicants for public housing in a given locality were permitted to designate into which project they wished to be placed; no tenant would be placed in a project that he or she found unacceptable.

The "free choice" plan completely ignored the reality of past discriminatory site selections in all-black and all-white neighborhoods. Free choice in the context of past governmentally sanctioned segregation was not free choice at all. Applicants viewed the projects as race-identified.

It was not until July 10, 1967, that HUD ordered all local housing authorities to put into effect a "first-come-first-served" program for tenant assignment. Under that plan, an applicant had to be offered vacancies in the locations having the most vacancies. If the applicant turned down a vacancy three times, his or her name dropped to the bottom of the list. This plan contained numerous exceptions, and in practice allowed the local authorities discretion to avoid instituting a plan that would in fact operate on a first-come first-served basis. In addition, it applied only to low-rent public housing rather than all subsidized rental housing.

The deficiencies of this laissez-faire approach are still being perpetuated in current HUD subsidized programs, notably the Section 8 Existing Housing Assistance Program.[91] Under the program, which began in mid-1975, local Public Housing Authorities subsidize rental housing by issuing certificates to families with incomes that are less than 80% of the area's median income. It is the responsibility of the certificate holder to find a unit that meets Section 8 standards and rents for less than the "Fair Market Rent," as calculated by HUD. The housing assistance payment is equal to the difference between the rent charged and 25% of family income adjusted for family size.

By 1978, over 133,840 units had been allocated. As of 1980, the program had become the nation's second largest low-income housing program, next to public housing, and as such had enormous potential as a vehicle for furthering residential integration. However, the laissez-faire underpinnings of the program and the failure of HUD to operate the program in an affirmative, integrative manner, left the residential market to operate as it would have absent the program.[92] Several evaluations confirm that housing opportunities have not been expanded under Section 8. In Winston-Salem, North Carolina, 70% of all blacks who received assistance under the Section 8 existing program remained in census tracts that were at least 90% black.[93] In Cuyahoga County (Cleveland), Ohio, a study found that 63% of those families who benefited from the rent subsidy program remained in the locations in which they had lived before receiving the subsidy. It indicated also that most white families searched for housing in their immediate neighborhoods, but that black families most often looked outside their immediate neighborhoods, possibly indicating that minorities were more dissatisfied than whites with the neighborhoods in which they lived. However, despite their desire to move, minority families were less likely than white families to locate suitable housing in a neighborhood that was more desirable than their current one (Battle et al., 1977).

Initially, the housing units eligible to be rented under the Section 8 Existing Program generally had to be located within the jurisdiction served by a local public housing agency. While HUD's equal opportunity regulations issued pursuant to Section 8[94] appear to require the PHA's to ensure that housing for eligible families can be found "in areas outside low-income and minority concentrations and outside the local jurisdiction where possible,"[95] PHA efforts to secure participation of owners outside such areas are required only "where possible." Thus blacks residing in the inner city more often than not go to the municipality's housing authority and are not given lists of housing in suburbs served by other housing authorities.[96] The consequence is that mobility is actually more limited than it would be in the private market.

The Section 8 program has also failed to promote integration because HUD has failed to require that tenants be counseled or otherwise assisted in attempts to find housing in neighborhoods where they would not be members of the predominant race of the residents. Such assistance, when actively given, as in Louisville and Jefferson County, Kentucky; Chicago, Illinois; and Milwaukee, Wisconsin, has in fact increased housing desegregation.

Finally, in all too many communities, HUD has not maintained records that would have enabled it to determine the placement of residents by race under the Section 8 Program, thereby making it impossible for HUD to monitor the racial impact of the Section 8 Program.

HUD'S FAILURE TO TAKE AFFIRMATIVE ACTION TO FURTHER FAIR HOUSING POLICIES

In 1968, Title VIII mandated that HUD "administer its programs and activities in a manner affirmatively to further the policies" of fair housing. By and large, HUD has been painfully slow, if not reluctant, to rise to that challenge.

HUD itself frequently has taken a restrictive view of its authority. In federal court cases, HUD argued for an extremely narrow interpretation of its obligation to act affirmatively to dismantle racial segregation, limiting its "affirmative duty" to a prohibition against funding local agencies that it knows to be practicing discrimination.[97] HUD failed to use its pre-grant review process and its compliance reviews to mandate the kind of aggressive, affirmative efforts necessary to undue the institutionalized dual market it was responsible for creating.

Again, HUD's administration of its Section 235 program in Kansas City, Missouri is typical of the administration nationally and is instructive here. The FHA's administration of its deep subsidy Section 235 program, from 1968 to 1973, caused white flight from the inner city, racial turnover, resegregation of white neighborhoods, and the rapid destabilization and resegregation of integrated neighborhoods.[98] FHA's home-buyer qualifications, its underwriting, appraisal, and inspection policies and practices, and its failure adequately to monitor the real estate and home finance industry actors that are subject to its authority[99] had many adverse effects. These include rapid turnover in home ownership, large inventories of vacant homes, plummeting property values in inner-city areas, disinvestment by lenders and insurers, neighborhood blight, and destruction of community social institutions.[100] Turnover in home ownership was sometimes more than ten times in ten years.

Typically, black families who had no preparation to become home owners were sold homes often in older, transitional neighborhoods. Many were unable to maintain their homes and abandoned them, or they were foreclosed. This accelerated the deterioration of newly integrated neighborhoods and reinforced the ghettoization process.[101] Vacant homes foreclosed by the FHA were not well maintained, and were often vandalized or burned out. Homes foreclosed under both the 203 and 235 programs often stayed vacant for long periods of time.[102] The FHA took the position that it was not responsible for the maintenance of a home until it actually acquired title.[103] As a result, a home often remained unattended from the time it was abandoned until the time the mortgage was actually foreclosed and turned over to the FHA, a period that often ran into months. Ironically, one of the appraisal criteria under HUD guidelines is the number of abandoned buildings in the neighborhood. One set of HUD policies thus reinforced and aggravated the effects of another.

From 1968 until 1973, during the years when the 235 program was most active and a tremendous amount of money was being fed into the housing market through this program, it was HUD's national practice not to monitor in any real way the racial residential effects of the program[104] or otherwise act affirmatively to counteract the momentum of private market forces.

Given the magnitude of the 235 program, that failure proved to be a large lost opportunity. Because of its significant subsidies the program was an extraordinarily attractive opportunity for a family to become a homeowner in a desirable area without having to put almost any money forward as a down payment and with low-cost monthly payments. Vigorous fair housing enforcement and housing counseling would certainly have found minority families who were interested in the new suburban housing and who would have been able to maintain home ownership.[105] Similarly, if HUD had been sensitive to the effect of its 235 program in older integrated neighborhoods, it could have limited that program in those neighborhoods. It could have provided counseling for the minority families about a wider range of housing opportunities and the responsibilities of home ownership. Home ownership counseling may have provided a reasonable chance of keeping their homes and avoiding abandonment and deterioration of neighborhoods.[106]

Housing counseling could have been decisive in the HUD-funded relocation programs as well: Counseling that encouraged white families to consider integrated neighborhoods as well as counseling for blacks as to integrated housing opportunities. These efforts could have been tied to an active fair housing enforcement effort to ensure that families using HUD relocation monies to find replacement housing were not subjects of private market discrimination. In communities where such counseling has been used, it has made a dramatic difference in where people resettle, as for instance, in the HUD-financed programs in Chicago, Louisville, and Baltimore. These programs actually take black families with housing subsidy payments to housing outside of traditional black areas and encourage them to apply for homes in those areas.[107]

The pattern of HUD's failure to act affirmatively is evident in its subsidized multifamily housing programs as well. The racial composition of HUD subsidized multifamily housing all too frequently mirrors the racial composition of the neighborhoods in which it is built. While this is not surprising, given the strength and momentum of private market forces, HUD could have done more to break these occupancy patterns. While it issued affirmative fair housing marketing regulations,[108] it failed effectively to monitor the affirmative marketing plan components and train the staff members who operate the housing units in methods of attracting and maintaining an interracial unit.[109] Having implemented regulations requiring developers to design affirmative marketing plans, HUD needed to enforce those plans to ensure that the plans would have an effect. All too often, there is little relationship between affirmative marketing plans and actual marketing practices by many developers because the developer's signing of an affirmative marketing plan has not been followed by any monitoring by HUD.[110]

Moreover, the kinds of affirmative steps required under the AHFM regulations, such as advertising in newspapers and publications with high black readership, has not proved as effective as fair housing counseling that actually takes minority families out to see subsidized projects in whites areas.[111]

Where HUD was forced under court order to change its housing policies in Chicago, Illinois, it has implemented affirmative programs that have been more effective in desegregation. In that city it contracted to provide housing counseling and created special allocations of subsidized housing units and certificates.[112] The result has been that despite that city's history of extremely segregated housing patterns and long tradition of racial hostility, many black families from largely black housing projects have moved into virtually all-white areas in the very distant suburbs of Chicago and have reported extremely positive experiences, which resulted in a tremendous demand for other black families to join in that process.[113]

The Section 8 program provides a final illustration of HUD's failure to act in a sufficiently affirmative manner in ensuring open housing. Regulations issued pursuant to Section 8[114] required the local housing authorities (PHAs) to ensure housing availability outside minority concentrations to assist certificate holders to find suitable housing when discrimination has prevented them from finding such housing. Yet the regulations fail affirmatively to attack the impediments of discrimination in that:

—PHA efforts to secure participation of owners outside areas of minority concentration are required only "where possible."
—PHAs are not instructed to investigate and resolve complaints of discrimination against owners who are participating in the rent subsidy program.
—PHAs are not instructed that, when they receive a complaint of discrimination that they cannot resolve voluntarily, they must refer that complaint to HUD or to a state or local agency with authority to resolve the matter.
—When an owner denies housing to a certificate holder because of race or sex, the PHA is not instructed to ensure that the owner be barred altogether from participation in the Section 8 rent subsidy program (U.S. Commission on Civil Rights, 1979).

Whether due to a lack of resources or imagination, HUD has failed at times to take affirmative steps that could counter the momentum of the many real estate market forces that perpetuate segregation.

HUD'S REFUSAL TO UTILIZE AVAILABLE SANCTIONS: SOME CASE STUDIES

Under the fair housing laws discussed in this chapter, HUD is responsible for ensuring equal housing opportunity in the assistance programs it operates. Yet HUD has not been forceful in monitoring and imposing sanctions to assure compliance.

In 1979 the United States Commission on Civil Rights found that

—HUD has established equal opportunity requirements, such as affirmative marketing plans, equal opportunity housing plans, and broker certifications, for many of its program participants, but it has not regularly monitored compliance with these requirements.
—HUD has conducted too few compliance reviews of recipients of HUD assistance. In fiscal year 1977, 21% of all compliance reviews of recipients focused on private sponsors and owners, representing less than 1% of these participants in HUD programs. In total, 56% of HUD's compliance

reviews focused on local public housing authorities, representing only 3% of these participants.

—In the preceding few years, HUD conducted only one communitywide pattern and practice investigation, under Title VIII, although in 1974 the Commission noted that conducting 50 such reviews in the next year was essential for meaningful Title VIII implementation.

—HUD's delays in complaint processing and its failure to use "testing" have curtailed the Department's ability to corroborate complainants' allegations of discrimination.

—HUD has not required prompt correction of noncompliance discovered through compliance reviews. It has been unwilling to terminate grant recipients upon a finding of civil rights violations, but instead has typically continued to carry out protracted negotiations beyond the 60-day limitation provided for in Departmental regulations (U.S. Commission on Civil Rights, 1979).

A classic example of HUD's apparent inability or unwillingness to impose sanctions on recipients of federal housing funds is its reaction to the Texarkana, Arkansas public housing authority (THA). Despite the local administration of the housing programs, HUD has retained a large amount of discretion to approve or reject both site selection and tenant assignment procedures of the local housing authority. Thus funding cutoffs would have been particularly valuable as a tool for ensuring fair housing compliance. Yet in Texarkana, Arkansas, despite its own repeated findings from 1969 to 1979 that the THA was not in compliance with Title VI, HUD refused to exercise its authority to cut off funding.

In a case brought against HUD for its handling of the blatant violations of housing laws in Texarkana, the U.S. Eighth Circuit Court of Appeals found in the case of *Client's Council v. Pierce*[115] that "for over a decade, HUD consistently responded to its own findings of noncompliance in ways that allowed the THA to continue discriminating. Suggestions of actions short of a funding cutoff were often made by HUD officials, but the agency ultimately rejected or ignored them, choosing instead to pursue a course of conduct that aggravated rather than alleviated the discrimination in Texarkana public housing."[116]

Faced with blatant segregation and an admitted determination to discriminate intentionally, HUD did nothing to change effectively the operation of the THA. In 1969, HUD made a finding of the local agency's noncompliance with Title VI and Title VIII but it was ignored by the THA and, after an exchange of letters, was forgotten by HUD.

Prodded into action by congressional inquires in 1971, HUD's equal opportunity staff found identical conditions of noncompliance. More letters were exchanged, a resolution was passed, and yet more funds were released. The pattern was repeated in 1972 and 1973.

After the HUD Regional Office concluded that the THA was still in noncompliance with Titles VI and VIII, that it was "nonresponsive" to any attempts to alleviate the discrimination, and that it was not even negotiating with HUD in good faith, these HUD Regional Office officials recommended that certain modernization funds be withheld from the THA. But HUD Area Office officials appealed to higher authorities in 1974, and HUD responded to their request by releasing the funds in question and by finding that the THA was in compliance with the law. This action restored the THA to a "normal posture" and allowed it to receive $1,475,528 in modernization funds and grants from 1974 through 1978 without restriction.

HUD officials who had been associated with the Texarkana case since 1969 vigorously protested the agency's decision to find the THA in compliance, but were overruled. At this point, even if the agency had determined that release of the modernization funds was warranted, alternatives were available. The agency could have released the modernization funds on the condition that the modernized units be filled on an integrated basis. The agency also could have referred the case to the Justice Department, as it ultimately did after the tenants initiated their suit. At the very least, HUD could have refused to find the local housing authority in compliance with Title VI, when it was obvious that blatant violations existed.

Conditions did not improve and, in 1978, HUD again found that the THA had failed to integrate a single project. Nevertheless, HUD simply recommended that the THA pass another resolution, even though the HUD official who conducted the review of the THA concluded that for nine years the THA's resolutions in response to HUD investigations had merely been a paper exercise. HUD's fair housing office took some positive action in 1978 by finding that the City of Texarkana was ineligible for a federal Urban Development Action Grant, in part because of the THA's segregated housing projects. Yet, even then, the various offices within HUD that were responsible for monitoring the THA's compliance with Title VI and VIII were divided and apparently immobilized. Although a HUD report in 1979 concluded that the THA had made substantial progress, HUD's fair housing office found that the THA remained in noncompliance. In any event, no effective sanctions were taken for the noncompliance before 1979, that is, for over a decade. The tenants eventually sued HUD in 1979, and the United States Court of Appeals ultimately found in their favor.[117]

Urban renewal programs administered by HUD were subject to similar deficiencies because of HUD's reliance on local agencies to administer them and its failure either to monitor them or to take swift and strong action to correct deficiencies in the local administration when they were found.

HUD's approach to noncompliance of local agencies administering the public housing and relocation projects in Independence, Missouri, a largely white community in the Kansas City, Missouri metropolitan area serves as another example.[118] Hocker Heights, a public housing project completed in 1965, was placed in one of the two substantially black neighborhoods in Independence; it displaced only black families from their homes, the vast majority of whom moved to the largely black inner city of Kansas City, Missouri.[119] About the same time, the oldest and best established integrated neighborhood in Independence was chosen as the site of the town's first major urban renewal project. While the urban renewal project, Northwest Parkway project, displaced whites as well as blacks from what was an integrated neighborhood, all the blacks in the project area and for several blocks on all sides of the area were uprooted, while most whites in the project were allowed to remain and rehabilitate their homes.[120] Moreover, the blacks who were displaced represented a much larger portion of the black population in Independence than did the white displacees relative to whites in Independence.[121] The displacement of a disproportionately large number of blacks from their homes in Independence had an indirect as well as a direct impact, discouraging other blacks from staying in or moving to Independence from the predominantly black sections of Kansas City, Missouri. Many blacks did not want to move to Independence because of its reputation for "Negro removal."[122]

Other subsequent urban renewal projects in Independence, Missouri were implemented despite the local urban renewal authority's own findings that virtually all blacks wanted to remain in Independence and that private housing in Independence was largely unavailable to blacks.[123] The local authority failed to consider the impact of the projects on blacks in that community. It turned a blind eye to discriminatory housing practices throughout the locality and failed to ensure adequate relocation facilities, substantial low-cost housing, and housing opportunities free of discriminatory practices.[124]

HUD not only received written complaints that discrimination was taking place, but HUD representatives actually visited Independence and spoke with black residents who described the hardships they faced, both in terms of this disproportionate displacement and their relocation in a highly discriminatory housing market.[125] Yet HUD refused to take any action that halted the projects or imposed sanctions on the Independence agency for continuing the projects even after 1964 when the continued funding of the projects arguably violated Title VI.[126]

The role played by HUD in the implementation of the public housing and urban renewal projects in Independence, Missouri was significant. The existence of those programs was entirely dependent upon continuing, sometimes year to year, federal financial assistance. HUD could have minimized the impact of the projects on black residents, including halting the dislocation of disproportionate numbers of blacks. It could have withheld funds or conditioned funding of the Independence Public Housing Authority on its allowing blacks who had relocated outside of Independence to return to live in the Hocker Heights public housing project.

The urban renewal programs in the Kansas City area continued to exacerbate racial segregation throughout the 1960s and into the 1970s. A similar pattern of noncompliance was taking place during the late 1960s in the urban renewal program in Kansas City, Missouri's inner city. For example, the urban renewal agency in Kansas City assigned its relocation specialists who aided displaced persons on the basis of race and it used separate lists of available houses for blacks and whites based on race.[127] It referred blacks to predominantly black neighborhoods and whites to predominantly white neighborhoods in the Kansas City area. The agency also regularly referred black relocates to black real estate agents and white relocatees to white agents.[128] Black relocates were more likely than were white relocates to be moved into substandard homes.[129] Homes in black areas that were demolished were undervalued, and black homeowners consequently received such low compensation that it was impossible to get a comparable home in areas outside those that were predominantly black.[130] Finally, in 1970, the agency still did not keep track of the referrals to replacement housing of displaced persons and HUD did not require it do so.

HUD's Fair Housing and Equal Opportunity Office made a compliance review of the Kansas City, Missouri urban renewal agency, the Land Clearance and Redevelopment Agency (LCRA) in 1971 and made several findings of blatant discrimination.[131] But HUD funding of the program was not cut off. Then from 1971 to at least 1975, HUD officials were repeatedly made aware of those practices and others that violated Title VI, including the continued practices by LCRA that resulted in relocation of blacks into black areas, and whites into white areas. Yet HUD failed to cut off federal funds or in any way bring significant sanctions to bear on the local agency until 1972, when, under threat of a cut off, the local agency underwent a major reorganization. However, by that time the damage had been done.[132]

According to the government's own figures, between 1963 and 1972 more than 2,532 households, 1,118 of them black, were relocated by federally funded urban renewal projects in the Kansas City metropolitan area.[113] In fact, these figures may be conservative because prior to 1971, a substantial portion, often more than half, of the families in designated urban renewal areas moved themselves without government aid in anticipation of being displaced, and these families went unrecorded by the government.[134] The total number of black households relocated in Kansas City, Missouri by both urban renewal and the Missouri State Highway Department was 6,213.[135] Moreover, that figure does not tell the whole story. To judge the full effect of relocation policies, the total number of white households—estimated to be 9,560[136]—must be considered because their movement to the white suburbs increased the racial segregation in the inner city.

The racial segregation resulting from the operation of HUD funded programs in Arkansas and Missouri was replicated in communities throughout the country. The massive dislocation of minority group populations funded with federal dollars and overseen by HUD continued and accelerated the growth of new segregated slums, thwarting the purpose for which urban renewal and related programs were enacted.

HUD'S TENDENCY TO BOW TO POLITICAL PRESSURES

As might be suspected, HUD is not immune to political pressure in the administration of its programs and that pressure is exerted both at HUD headquarters in Washington, D.C. as well as in its regional and area offices. At times that pressure results in decisions that work against the agency's stated goal of fair housing. For example, in some cases, despite HUD's site selection criteria, it has explicitly bowed to local interest groups with agendas opposed to integration. In Chicago, Illinois, for example, HUD approved and funded regular family housing sites chosen by the Chicago Housing Authority (CHA) between

1950 and 1969, knowing that such sites were not "optimal" and that the reason for their exclusive location in black areas of Chicago was that "sites other than in the south or west side, if proposed for regular family housing, invariably encontered sufficient opposition in the [Chicago City] Council to preclude Council approval."[137] A similar pattern was evident in the Philadelphia, Pennsylvania public housing program.[138]

HUD's actions in disapproving projects when faced with public opposition, after substantial investment by developers, had another negative effect in that they discouraged many developers and architects from becoming involved in the construction of federally subsidized, low-income housing. The ability to overcome neighborhood opposition to subsidized housing should begin with education of local residents, and HUD could have taken an active role, but in most cases it neither required the persons or entity receiving funds to do so nor did so itself.

The fervor with which HUD has embarked on affirmative programs designed to enforce and promote fair housing has also varied, depending on the particular administration in power in Washington. Some observers have noted that the leadership provided by Presidents Nixon and Ford in this area was unremarkable for the most part, and that while the Carter administration took a more active stance on fair housing issues, it too was not sufficiently aggressive (Lamb, 1981).

Under the Reagan administration, some proponents of fair housing argue, the federal government has actually been in a period of retrenchment, substantially reducing activities to enforce open housing, and even dismantling some of the programmatic structures upon which efforts toward fair housing had been built (Sloane, 1983; Weaver, 1985). For instance, following the Reagan administration's changes in the U.S. Commission on Civil Rights membership, that body has not produced any progress reports on fair housing enforcement of the kind that it had authored on a regular basis since it was established in the late 1950s. Significantly, although President Reagan's Commission on Housing, established by Executive Order on June 16, 1981, was called upon to develop a national housing policy and advise the president "on the role and objectives of the Federal government in the future of housing" (U.S. President's Commission on Housing, 1982), neither the Commission's mandate from President Reagan nor its resulting report addressed the problems and challenges of residential racial segregation facing the nation (U.S. President's Commission on Housing, 1982). Changing political orientations are designed to and do affect the fair housing enforcement efforts at the programmatic level.

HUD'S BUDGETARY AND STAFFING LIMITATIONS AND CHOICE OF PRIORITIES

The low priority that has been accorded to HUD's administration of Title VIII is evident from its staffing patterns, budget allocations, and organizational structure. In fiscal year 1978, for instance, HUD used only a small portion of its staff resources for Title VIII duties in the regions where most HUD Title VIII compliance activities take place, and allocated only about $5.8 million for Title VIII activities in all HUD offices. Such resources are inadequate if HUD is to carry out such activities as communitywide pattern and practice reviews and a comprehensive program of leadership and guidance for other federal agencies with Title VIII responsibilities (U.S. Commission on Civil Rights, 1979).

Moreover, the staff that HUD did have was not always organized for its most effective use. For instance, HUD failed internally to integrate equal housing opportunity concerns into its programmatic operations. Equal housing opportunity requirements were monitored by a separate staff responsible for measuring compliance with Title VI and Title VIII although it would have been far more effective if the same staff that managed the actual programs, whether it was Section 235 housing, public housing, or Section 8 subsidies, was responsible for and committed to running the programs with methods designed to dismantle segregative patterns.

CONCLUSION

To the same extent that the federal government acted to adopt and reinforce the segregative practices of the private real estate market, the federal government has an obligation affirmatively to dismantle residential segregation. It is a duty embodied in the fair housing laws of the 1960s, yet the promise those laws held forth remains unfulfilled.

Any successful change in the complexion of residential neighborhoods would require altering the practices of real estate agents and brokers, mortgage lenders, insurers, appraisers, developers—in sum, all the actors and institutions in the housing market. The deeply entrenched segregative behavior on the part of these actors, which perpetuates the dual housing market, would be extremely difficult to effect, even with the strongest of intentions and legal or fiscal muscle. The federal government, especially HUD, was perhaps the most natural entity to flex such muscle on a national scale. During the preceding decades, when federal dollars financed or insured new construction and rehabilitation projects across the country on a major scale, HUD could have attached conditions to its purse strings that would have necessitated a change in the behavior of actors in the housing market. Appraisers, for example, could have been required to demonstrate that they put a premium on houses in integrated neighborhoods; brokers and agents could have been forced, as a condition of doing business with the federal government, to demonstrate that black home seekers and renters were being shown homes in predominantly white neighborhoods; and insurers could have been required to insure inner-city homes. The measure of compliance should have been results. Indeed, HUD, with its own staff, could have done far more to educate home seekers, to attract minority applicants for subsidized housing to predominantly white neighborhoods, and vice versa.

Unfortunately, the federal agency did not exercise what statutory and regulatory power it had with the aggression and determination necessary to change housing market patterns. It failed to allocate sufficient staff and resources to promote integration or failed generally to interpret its authority in a broad manner. Now, with federally assisted housing construction at far lower levels than in the past, the opportunity for the federal government to dismantle the dual, segregative housing market is, as a practical matter, considerably less; the federal fiscal muscle has been considerably weakened.

To be sure, the process of ensuring open housing implicates political as well as legal and programmatic considerations. Unfortunately the nation no longer looks to the federal government for tough new action. Indeed, in the past few years, the federal government has backed away from tapping what authority it could wield by effectively interpreting and enforcing federal laws now on the books.

Proponents of fair housing must seek redress not only in the legal arena, by lobbying for tougher laws at the federal and at local levels and by instituting individual suits where discrimination arises, but also in the political arena by exerting pressures at the federal and local levels for a programmatic agenda that takes seriously the mandates embodied in the fair housing initiatives of the 1960s.

NOTES

1. The Department of Housing and Urban Development became a department in 1965. Its predecessor agency was the Housing and Home Finance Administration

(HHFA), established in 1937 to administer the United States Housing Act of 1937, 50 Stat. 888. The functions of HHFA, which included within it the Urban Renewal Administration (URA), the Community Facilities Administration (CFA), the Federal National Mortgage Association (FNMA), the Public Housing Authorities (PHA), the Federal Housing Administration (FHA), and other federal housing-related programs, were transferred to HUD in 1965. 79 Stat. 667 (1965).

2. The housing market is the production, distribution, consumption, and regulation of housing. A dual housing market exists where there are two separate markets, as defined above—one for blacks and one for whites. They intersect only in transitional neighborhoods where the race of the residents is changing from white to black.

3. While this chapter focuses on a number of federally subsidized programs, it does not purport to be an exhaustive treatment of the programs by which the dual housing market benefited from federal government largesse. Other forms include creation in 1932 of the Federal Home Loan Bank System to provide assistance to major home financing institutions; authorization of savings and loan charters; the 1938 creation of the Federal National Mortgage Association (FNMA) to provide a secondary market for home loans; loans and grants for water and sewer systems and other municipal facilities; and, more recently, block grants to local governments for community development.

4. For the proposition that residential segregation is of comparatively recent origin, dating to roughly the 1910s, see Weaver, 1983, pp. 1-2.

5. The Federal Housing Administration, established by the Housing Act of 1934, was created to insure residential mortgages on private dwellings against the failure of the individual borrower to repay and did so under the "Section 203" program. FHA Underwriting Manual, Section 101 (1935). The FHA's mortgage insurance eliminated lending institutions' risk in providing mortgage financing for properties meeting FHA standards of housing quality and appraised market value, thus encouraging lenders to accept lower interest rates and much longer repayment periods.

6. *FHA Underwriting Manual* Sec. 980 (3) (g) (1938).

7. 334 U.S. 1 (1948). In *Shelley v. Kraemer* the Supreme Court ruled that covenants by race and religion were legally unenforceable in the courts but the Supreme Court did not outlaw their voluntary use.

8. *FHA Underwriting Manual*, Part II, Sec. 310 (1935).

9. *FHA Underwriting Manual* Sec. 226 (November 1, 1936).

10. *FHA Underwriting Manual* Sec. 1320 (1958).

11. In Kansas City, Missouri, for example, the Kansas City Urban League estimated that in the entire period from 1925 to 1950, less than 300 new homes were made available to blacks throughout that city's residential market. Testimony of Dr. Gary Orfield, in the case of *Jenkins v. Missouri* (No. 77-420-Cv-V [W.D. Missouri]) at 14,846.

12. At the same time, federal income tax policies further promoted this phenomenon of housing segregation because deductions for federal income tax purposes of real estate taxes and mortgage interest significantly eased the financial burdens of suburban homeowners.

13. Testimony of Gary Orfield, February 29, 1984, in *Jenkins v. Missouri* at 14,855-56. The Kansas City school desegregation case of *Jenkins v. Missouri* (77-420 CV-W, W.D. Mo.) cited throughout this chapter was filed on May 26, 1977, and was tried between October 31, 1983, and Spring of 1984. The trial court found Kansas City Missouri School District and the State of Missouri liable and the U.S. Department of HUD not liable for

the unconstitutional segregation of the Kansas City Public Schools, *Jenkins v. Missouri* 593 F. Supp. 1485 (W.D. Mo. 1984) (the court dismissed the suburban school districts, and the Department of Health, Education and Welfare prior to the end of trial).

Citations to testimony in *Jenkins v. Missouri* hereinafter will be to the witness, date of testimony, and page of the Court reporter's official transcript.

14. U.S. Commission on Civil Rights (1974) at 46. See also, Grier and Grier, 1966, p. 43. "Often [the expressways being planned or under way in many cities were] designed to pass through the heart of the Negro 'ghettos.' Because of blight, property in these areas [could] usually be acquired under condemnation proceedings at relatively low costs. In addition, the federal program subsidizing expressway construction offered a convenient means by which to demolish unslightly slums at minimum cost to the local community."

15. Broad-based, low-income public housing was first funded by the United States Housing Act of 1937 that was amended by the Comprehensive Housing Act of 1949, P.L. 171, 63 Stat. 413. The public housing program provided for the federal government to furnish loans and financial assistance to local housing authorities that planned, built, and managed the projects subject only to overall federal policies and regulations.

16. For example, the Kansas City Housing Authority segregated all its projects by race until at least 1960, and thereafter at least some of its projects remained segregated until the 1970s.

17. See discussion of the Texarkana Public Housing Authority infra.

18. See testimony of Galen Bridges (former Director of Public Housing Authority of Kansas City, Missouri), in *Jenkins v. Missouri*, Deposition at 58-59, 62, 99 (discussing the interaction, in a segregative way, between public housing and urban renewal).

19. In Kansas City, Missouri all public housing projects were located in the inner city until as late as 1966.

For a discussion of the battles waged over public housing site selections since shortly after World War II, see Forman, 1971; Freedman, 1969.

20. The analysis of the statutes discussed in this section is taken largely from the U.S. Commission on Civil Rights, The Federal Fair Housing Enforcement Effort (1979) at 1-4. Judicial initiatives on a parallel track complemented the legislative ones. For example, *Jones v. Mayer*, 392 U.S. 409 (1968); (the Thirteenth Amendment of the United States constitution serves as a constitutional basis for legislation designed to prohibit racial discrimination in housing); *Reitman v. Mulkey*, 387 U.S. 369 (1967) (state constitutional amendment giving individuals discretion to discriminate in the sale or rental of property held unconstitutional as a violation of the Equal Protection Clause of the U.S. Constitution).

21. 24 CFR Part 107, published 45 FR 59514, Sept. 9, 1980.

22. 24 CFR Sec. 107.15. See discussion of the affirmative fair housing marketing requirements and Title VIII infra.

23. 24 CFR Sec. 107.35.

24. 24 CFR Sec. 107.40 et seq.

25. 24 CFR Sec. 107.50.

26. 24 CFR Sec. 107.60.

27. 24 CFR Sec. 107.65.

28. 42 U.S.C. Secs 2000d-2000d-6 (1970). Title VI provides that, "No person in the United States shall, on the ground of race, color, or national origin, be excluded from participation in, be denied the benefits of, or be subjected to discrimination under any

program or activity receiving Federal financial assistance." P.L. 88-352, 78 Stat. 252 (July 2, 1964).

29. 42 U.S.C. Sec. 2000d-4 (1970) provides: "Nothing in this [title] shall add to or detract from any existing authority with respect to any program or activity under which federal financial assistance is extended by way of a contract of insurance or guaranty.

30. 42 U.S.C. Sec. 2000d-1 (1970).

31. See the Attorney General's "Guidelines for the Enforcement of Title VI, Civil Rights Act of 1964," 28 C.F.R. Sec. 50.3 (1977).

32. Those regulations, issued in December 1964, have remained substantially unchanged to the present. Changes worthy of note for this discussion have been noted in the discussion that follows.

33. 24 CFR Sec. 1.4. In 1967, the Title VI regulation's proscription against specific discriminatory actions was expanded to include a provision requiring recipients operating federally assisted low-rent housing to assign residents on a communitywide basis in accordance with a HUD approved plan that would take into account factors that would promote the goals of Title VI. 24 C.F.R. Sec. 1.4(b)(2)(ii) (1985) (first published as a Final Rule at 32 F.R. 14819, Oct. 26, 1967).

By 1973, HUD additionally authorized HUD officials to prescribe plans for the assignment and reassignment of eligible tenants in federally assisted low-rent housing. 24 CFR Sec. 1.4(b)(2)(iii) (1985).

By 1973, HUD had also promulgated regulations concerning site selection that prohibited the recipients from making selections that had discriminatory purpose or effect. 24 CFR Sec. 1.4(b)(3) (1985).

By 1973, HUD had added to the proscriptions against discrimination the denial of the opportunity to participate as a member of a planning or advisory body integral to a program. 24 CFR Sec. 1.4(b)(1)(vii) (1985) (first published as a Final Rules at 38 FR 17949, July 5, 1973).

34. 24 CFR Sec. 1.5 (1965).

35. 24 CFR Sec. 1.6(b) (1965).

36. 24 CFR Sec. 1.7(a) (1965).

37. 24 CFR Sec. 1.7(b) (1965).

38. 24 CFR Sec. 1.7(6) (1985). See 24 CFR Sec. 1.7(b) (published at 38 FR 17949, July 5, 1973).

39. 24 CFR Sec. 1.4(b)(6)(i)

40. 24 CFR Sec. 1.4(b)(6)(ii).

41. 24 CFR Sec. 1.5(a)(2).

42. 42 U.S.C. Secs 3601-19, 3631 (1970 and Supp. 1986). P.L. 90-284, 82 Stat. 84 (April 11, 1968.)

43. Exempted from Title VIII are single-family homes sold or rented without the use of a broker and without discriminatory advertising; rooms or units in dwellings with living quarters for no more than four families, provided that the owner lives in one of them and does not advertise or use a broker. 42 U.S.C. Sec. 3603(b)(1970). In addition, religious organizations and affiliated associations are free to give preference in selling or leasing housing to persons of the same religion, provided that the property is not owned or operated for a commercial purpose and provided that the religion itself does not restrict membership on account of race, color, sex, or national origin. Private clubs and religious organizations that are not open to the public and which incidentally operate noncommer-

cial housing may limit occupancy of the housing to their members. (42 U.S.C. Sec. 3607) (1977).

44. 42 U.S.C. Sec. 3604(a) (Supp. V 1975).

45. 42 U.S.C. Sec. 3604(b)(Supp. V 1975); See, for example, *Williams v. Hamptom Management*, 339 F. Supp. 1146 (N.D. Ill. 1972); *Smith v. Soid Adler*, 436 F. 2d 344 (7th Cir. 1970).

46. 42 U.S.C. Sec. 3604(c) (Supp. V 1975).

47. 42 U.S.C. Sec. 3604(d) (Supp. V 1975).

48. 42 U.S.C. Sec. 3605 (Supp. V 1975).

49. 42 U.S.C. Sec. 3605 (Supp. V 1975).

50. 42 U.S.C. Sec. 3606 (Supp. V 1975).

51. 42 U.S.C. Sec. 3604(e) (Supp. V 1975).

52. 42 U.S.C. Sec. 3605 (Supp. V 1975).

53. 42 U.S.C. Sec. 3608(c) and (b) (1970).

54. 42 U.S.C. Sec. 3608 (d) (5). In addition, Section 808(d) requires all executive departments and agencies to administer "their programs and activities relating to housing and urban development in a manner affirmatively to further the purposes of this title," and "cooperate with the Secretary of HUD to further such purposes." 42 U.S.C. Sec. 3608(c) (1970).

55. 42 U.S.C. Sec. 3613 (1970).

56. 42 U.S.C. Sec. 3612.

57. 42 U.S.C. Sec. 3612. For a discussion of the efficacy of the Title VIII administrative process for redressing complaints of housing discrimination see generally Chandler, 1973; Comment, 1975.

58. 42 U.S.C. Sec. 3614.

59. Pub. L. 98-620, Title IV, Sec. 402(4), Nov. 8, 1984, 98 Stat. 3360.

60. 24 CFR Part 71 (1970).

61. 24 CFR Sec. 71.1 (1970). See 24 CFR Sec. 200.300 (published at 36 FR 24467, December 22, 1971), which prohibits discriminatory practices by any entity receiving benefits of FHA mortgage insurance or doing business with the FHA. 24 CFR Sec. 200.315.

62. 24 CFR Sec. 71.18-71.20.

63. 24 CFR Sec. 71.31 (1970).

64. 24 CFR Sec. 71.32.

65. See U.S. Commission on Civil Rights, 1973, at 110-14 for a full discussion of the problems that attend focusing exclusively on the complaint process.

66. 24 CFR Sec. 200.600, issued at 37 FR 75, January 5, 1972, pursuant to EO. 11063 and Title VIII, and amended, 40 FR 20080, May 8, 1975.

67. Although the Federal Housing Administration (FHA) no longer exists, HUD continues to use the term FHA to describe the programs once administered by that agency.

68. 24 CFR Sec. 200.610 (1985).

69. 24 CFR Sec. 200.610 (a)-(f).

70. 24 CFR Sec. 200.635 (1977). HUD Form 935.2 (1976).

71. 24 CFR Part 108, (published 44 FR 47013, Aug. 1979, amended in part 50 FR 9168, March 7, 1985.)

72. 24 CFR Sec. 108.1.

73. 24 CFR Sec. 108.20.

74. 24 CFR Sec. 108.21.

75. 24 CFR Sec. 111.101. The Fair Housing Assistance Program was implemented in 1980 (45 FR 31881, May 14, 1980).

76. A recent evaluation of the Fair Housing Assistance Program found generally favorable results. Local agencies were reported to be handling a larger number of fair housing cases, in a more timely manner and with better settlements for successful complainants, while expanding fair housing activities such as outreach, public education, affirmative marketing, and testing. However, that report also cited shortcomings including coordination problems within HUD's Office of Fair Housing and Equal Opportunity that impeded program implementation (Wallace et al., 1985).

77. 24 CFR Part 109, (published 45 FR 57105, Aug. 26, 1980).

78. 42 USC 3604(c).

79. 24 CFR Sec. 109.16.

80. CFR Sec. 109.30.

81. 15 U.S.C. Sec. 1691 et seq.

82. 12 U.S.C. Sec. 2901 et seq.

83. For instance, the overall responsibility for prescribing regulations to carry out the purpose of the Equal Credit Opportunity Act is assigned to the Federal Reserve Board. 15 U.S.C. Sec. 16916, (1976).

84. The record of the Justice Department in prosecuting federal claims has varied. In the years immediately following the passage of Title VIII, the department vigorously pursued a civil rights housing docket and was responsible in significant part for advocating the expansive interpretation of Title VIII that the federal courts have taken, for example, *U.S. v. Mitchell*, 327 F. Supp. 476 (N.D. Ga. 1971) (panic selling); *United States v. American Institute of Real Estate Appraisers*, 442 F. Supp. 1072 (N.D. Ill. 1977) (discrimination in foreclosure and appraisal practices); *United States v. City of Parma*, 494 F. Supp. 1049 (N.D. Ohio 1980), affirmed in part 661 F.2d 562(6th Cir. 1981) (broad scale municipal policies on housing). However, under the Reagan administration, such vigorous efforts have all but ceased. For more than one year after the Reagan administration took office, the Department of Justice did not file a single fair housing lawsuit; two years thereafter, the department had filed a total of six fair housing lawsuits, only one of which has the potential for establishing significant legal precedent or being about some kind of institutional reform (Sloan, 1983).

85. See Sen. Rep. No. 96-919, 96th Cong. lst Session (August 26, 1980) at 20-26.

86. For example, it was not until the early 1970s that HUD even kept track of the race of the beneficiaries of its programs. By regulation promulgated in 1971, HUD required all participants in HUD programs to furnish information concerning minority-group identification to assist HUD in carrying out its responsibility for administering the national policies prohibiting discrimination and for fair housing under E.O. 11063, and Titles VI and VIII. 24 CFR Sec. 100.2, published 36 F.R. 24458, December 22, 1971.

87. Testimony of Gary Orfield, February 27, 1984, in *Jenkins v. Missouri* at 15,212.

88. See also testimony of Robert Lee Newsome, January 16, 1984, *Jenkins v. Missouri* at 9438.

89. Testimony of Gary Orfield, February 27, 1984, in *Jenkins v. Missouri* at 15,209 and 15,213.

90. The FHA Section 235 program was enacted in 1968. The Housing and Urban Development Act of 1968, Secs. 101(a), 201(a), P.L. 90-488 (Aug. 1, 1968), added Sec. 235 to the National Housing Act. Under that program, HUD made monthly payments to a

commercial mortgage lender to reduce the borrower's interest payments to as low as 1% with minimal down payments.

91. Authorized by the Housing and Community Development Act of 1974, P.L. 93-383, 88 Stat. 633.

92. A study by the Joint Committee on Public Housing in Kansas City, Missouri, for example, found that in December 1976, Section 8 units had frequently been located primarily in or adjacent to the racially effected neighborhoods. Officials administering the Section 8 program in Kansas City concluded that since the Section 8 program depended on the willingness of private owners or rental agents to rent to blacks, all the discriminatory forces at work in the private housing market served to prevent blacks from living outside the area of black concentration.

93. HUD Grantee Performance Report, Winston-Salem, North Carolina (April 1977).

94. 24 CFR 882.204(b) (1977).

95. 24 CFR 882.204(b) (1977).

96. At least one court has found that the freedom of choice aspect of Section VIII and the failure to operate the program on a metropolitanwide basis is illegal where to do so perpetuates segregation. *Jaimes v. Toledo Metropolitan Housing Authority*, 758 F.2d 1086 (6th Cir. 1985), but the operation of the program on a national basis has remained unchanged.

97. See, for example, *Clients' Counsel v. Pierce*, 711 F.2d 1406 (8th Cir. 1983).

98. Testimony by offer of proof of Dr. James A. Kushner, January 31, 1984 in *Jenkins v. Missouri* at 53-54; Testimony of Dr. Gary Orfield, February 28, 1984 in *Jenkins v Missouri* at 15,195-198, 15,642 et seq.; Testimony of Richard H. Fischman, February 2, 1984 in *Jenkins v. Missouri* at 12,129-12,303; Testimony of Charles Hammer, February 1, 1983, in *Jenkins v. Missouri* at 11,994 et seq.

99. Testimony by offer of proof of James A. Kushner, January 31, 1984, at 8-31.

100. Testimony of Gary Tobin, February 9, 1984, in *Jenkins v. Missouri* at 13,240-13,242. Testimony by offer of proof of Gary Orfield, February 29, 1984, in *Jenkins v. Missouri* at 28-32, 38-39.

101. Testimony on Albert Byrd, November 10, 1983, in *Jenkins v. Missouri* at 2321.

102. Testimony of Father Norman Rotert, January 24, 1984, in *Jenkins v. Missouri* at 10,722-10,724.

103. Testimony by offer of proof by James A. Kushner, January 31, 1984, in *Jenkins v. Missouri* at 8-11.

104. Testimony of Gary Orfield, February 27, 1984, in *Jenkins v. Missouri* at 15,244.

105. Testimony of Gary Orfield, February 27, 1984, in *Jenkins v. Missouri* at 15,246.

106. Testimony of Gary Orfield, February 27, 1984, in *Jenkins v. Missouri* at 15,246-47.

107. Testimony of Gary Orfield, February 27, 1984, in *Jenkins v. Missouri* at 15,265-15,266.

In contrast, one of the early counseling services established in Indianapolis during the mid-1960s, which sought to coordinate with the FHA Counseling Service in that city, took a more race-neutral and less affirmative approach than that proposed by Dr. Orfield. The account of that effort at open housing described in Baum (1971) is especially instructive as to the hurdles set up by both certain segments of the FHA and private real estate institutions.

108. 24 CFR Sec. 200.600, et seq.

109. Testimony of Gary Orfield, February 27, 1984, in *Jenkins v. Missouri* at 15,271-15,272.

110. Testimony of Gary Orfield, February 28, 1984, in *Jenkins v. Missouri* at 15,645.

111. Testimony of Gary Orfield, February 28, 1984, in *Jenkins v. Missouri*, at 15,515.

112. Testimony of Gary Orfield, February 28, 1984, in *Jenkins v. Missouri*, at 15,637.

113. Testimony of Gary Orfield, February 28, 1984, in *Jenkins v. Missouri*, at 15,635-36.

114. 24 CFR Sec. 882.204(b) (1977).

115. 711 F.2d 1406 (8th Cir. 1983).

116. The discussion of the facts that follows is taken directly from the Appeals Court's findings.

117. Client's Council v. Pierce, 711 F.2d 1406 (8th Cir. 1983)

118. The discussion of facts that follows is based on evidence offered by plaintiffs at trial of the Kansas City, Missouri school desegregation case of *Jenkins v. Missouri* (77-420 CV-W, W.D. Mo.). After hearing this evidence, the trial court did not find that HUD was liable under the Constitution, Title VI and Title VIII for the segregative impact of the Independence and Kansas City, Missouri urban renewal and housing projects discussed herein.

119. Testimony of Nathaniel Moreland, November 28, 1983, in *Jenkins v. Missouri* at 3737, 3767-3804.

120. Testimony of Nathaniel Moreland, November 28, 1983, in *Jenkins v. Missouri*. See also Testimony of Reva Fender, February 7, 1984, in *Jenkins v. Missouri* at 12778 et seq.

121. Testimony of Reva Fender, February 7, 1984, in *Jenkins v. Missouri* at 12778 et seq. See also Plaintiffs' trial exhibit 393 (Complaint Filed Against the City of Independence).

122. Testimony of Reva Fender, February 7, 1984, in *Jenkins v. Missouri* at 23778 et seq; Testimony of Gwendolyn Wells in *Jenkins v. Missouri* at 6685-6694.

123. Plaintiffs' trial exhibit 670 (Independence LCRA Report on Local Housing Supply) in *Jenkins v. Missouri*.

124. Testimony of Reva Fender, February 7, 1984, in *Jenkins v. Missouri* at 12778 et seq. and Nathaniel Moreland, November 28, 1983, in *Jenkins v. Missouri* at 3737, 3767-3804.

125. Testimony of Reva Fender, February 7, 1984, in *Jenkins v. Missouri* at 12778 et seq.

126. Testimony of Reva Fender, February 7, 1984, in *Jenkins v. Missouri* at 12778 et seq.

127. Testimony of Joyce Keller, former employee of the Kansas City, Missouri, LCRA, February 6, 1984, in *Jenkins v. Missouri* at 12429 et seq.

128. Testimony of Joyce Keller, former employee of the Kansas City, Missouri, LCRA, February 6, 1984, in *Jenkins v. Missouri* at 12429 et seq.

129. Testimony of Joyce Keller, former employee of the Kansas City, Missouri, LCRA, February 6, 1984, in *Jenkins v. Missouri* at 12429 et seq.

130. Testimony by offer of proof of James Kushner, January 31, 1984, in *Jenkins v. Missouri* at 45-48.

131. Plaintiff's trial exhibit No. 2659 (HUD's Final Investigative Report on LCRA, March 20, 1972) in *Jenkins v. Missouri*.

132. Testimony of Rafael San Juan, Kansas City, Missouri, January 30, 1984, at the trial of *Jenkins v. Missouri* at 11508, 11,536-11,584.

133. Plaintiff's exhibit number 353 F (Summary of Households Relocated from Federally Funded Projects Causing Displacement) admitted into evidence in *Jenkins v. Missouri* at 10,930.

134. Testimony of Yale Rabin, January 24, 1984, in *Jenkins v. Missouri* at 10,930 -10,931.

135. Plaintiffs' Exhibit number 353I (Consolidated Estimate of Displacements and Relocations from Urban Renewal and Highway Rights of Way by Year) admitted into evidence in *Jenkins v. Missouri.*

136. Plaintiffs' Exhibit number 353I (Consolidated Estimate of Displacements and Relocations from Urban Renewal and Highway Rights of Way by Year) admitted into evidence in *Jenkins v. Missouri.*

137. *Gautreaux v. Romney*, 448 F.2d. 731, 737 (7th Cir. 1971).

138. *Residents Advisory Board v. Rizzo*, 564 F2d 126 (3d Cir. 1977), cert. denied, sub. nom, *Whitman Area Improvement Counsel v. Residents Advisory Board*, 98 S.Ct. 1457 (1979).

REFERENCES

Joseph H. Battle and Associates. (1977) *The Section 8 program for existing housing in Cuyahoga County Shaker Heights, OH: Author*

Baum, D. J. (1971). Toward a free housing market. Coral Gables, FL: University of Miami Press.

Chandler, J. P. (1973). Fair housing laws: A critique. *Hasting Law Journal, 24*, 159.

Comment. (1969). The federal fair housing requirements, Title VIII of the 1968 Civil Rights Act. *Duke Law Journal, 733*, 748-749.

Comment. (1975). The Fair Housing Act of 1968: Its success and failure. *Suffolk University Law Review, 9*, 1312.

Falk, D., & Franklin, H. (1976.), *Equal housing opportunity: The unfinished agenda.* Washington, DC: Potomac Institute.

Forman, R. E. (1971). *Black ghettos, white ghettos and slums.* Englewood Cliffs, NJ: Prentice-Hall.

Freedman, L. (1969). *Public housing: The politics of poverty.* New York: Holt, Rinehard & Winston.

FHA. (1960a). The FHA and its system of mortgage insurance. (pp. 1-7). In *FHA underwriting training handbook* (vol. I.).

FHA. (1960b). Analysis of risks involved in mortgage transactions. In *FHA underwriting training handbook* (Vol. I). P. 2). Washington, DC: Government Printing Office.

Grier, G., & Grier, E. (1966). *Equality and beyond: Housing segregation and the goals of the great society.* Chicago: University of Chicago Press.

Lamb, C. (1981). Housing discrimination and segregation in America: Problematical dimensions and the federal legal response. *Catholic University Law Review, 30*(363), 405-406.

Orfield, G. (1974). Federal policy, local power and metropolitan segregation. *Political Science Quarterly, 89*, 777, 785.

Romney, G. (1970). *Equal educational opportunity: Hearings before the Senate select committee on equal educational opportunity* (91st Cong. 2d Sess). Washington, DC: Government Printing Office.

Sloan, M. (1983). Federal housing policy and equal opportunity. In *A sheltered crisis: The state of fair housing in the 80's*. Washington, DC: Commission on Civil Rights.
Taeuber, K. (1983). Racial residential segregation: 1980. In Citizens' Commission on Civil Rights, *A decent home for all* (Appendix).
U.S. Commission on Civil Rights. (1971). *A report of the racial and ethnic impact of the Section 235 program*. Washington, DC: Government Printing Office.
U.S. Commission on Civil Rights. (1973). *Understanding fair housing*. Washington, DC: Government Printing Office.
U.S. Commission on Civil Rights. (1974). *The federal civil rights enforcement effort: 1974* (Vol. II). Washington, DC: Government Printing Office.
U.S. Commission on Civil Rights. (1975). *Twenty years after Brown: Equal opportunity in housing*. Washington, DC: Government Printing Office.
U.S. Department of Housing and Urban Development. (1973). Historical overview: Equal opportunity in housing. In *Equal opportunity in housing reporter* (p. 2320). Englewood Cliffs, NJ: Prentice-Hall.
U.S. Department of Housing and Urban Development. (1976). *Affirmative fair housing marketing techniques: Final project* (Vol. I). Washington, DC: Office of Policy Development and Rcscarch.
U.S. President's Commission on Housing. (1982). *Report*. Washington, DC: Government Printing Office.
Wallace et al. (1975). *The fair housing assistance program evaluation* (prepared for HUD's Office of Policy Development and Research). Cambridge, MA: Abt.
Weaver, R. C. (1983). Housing discrimination: An overview. In *A sheltered crisis: The state of fair housing in the eighties* (presentations at a consultation sponsored by the U.S. Commission on Civil Rights) (pp. 1-6). Washington, DC: Commission on Civil Rights.
Weaver, R. C. (1985). Fair housing: The federal retreat. *Journal of Housing, 42*, 885-887.

EPILOGUE

The Costs of Housing Discrimination and Segregation: An Interdisciplinary Social Science Statement

PREFACE

This statement grew out of a meeting at the University of Chicago in February 1986, organized by Gary Orfield and the Leadership Council for Metropolitan Open Communities of Chicago. The statement has been signed by more than 25 scholars for a variety of disciplines whose work has addressed the issue of discrimination and segregation in housing. For a list of these signatories, contact the Leadership Council for Metropolitan Open Communities, 401 South State Street, Suite 860, Chicago, IL 60605.

I. INTRODUCTION

Housing discrimination is generally cast as a problem of unfair treatment of individual black (and other minority) homeseekers. Fair housing laws address this problem and surveys show that the great majority of Americans believe it has been largely solved.

It has not been solved. Urban racial ghettos persist in all of our great cities and indeed continue to expand. Both the residential segregation that ghettos embody and the system that perpetuates ghettos have effects that go far beyond the treatment of individual minority households. Racial ghettos have become one basic structural feature of American urban society.

The ghetto system exacts costs on individuals, neighborhoods, and the entire metropolitan structure. It shapes human relationships among major social groups. And it penalizes owners, investors, and job seekers.

No comprehensive research program has ever been launched to assess the full costs of segregation and discrimination. But certain findings within individual disciplines answer some key questions. In order to summarize existing knowledge and to chart promising avenues for future research, a meeting of researchers was held in Chicago in February 1986. Jointly organized by Gary

Orfield and the Leadership Council for Metropolitan Open Communities, the meeting included students of housing segregation and discrimination from the disciplines of economics, political science, geography, sociology, public policy, demography, urban planning, and history.

This statement is the outcome of that meeting. In its final form, it presents the results of further discussion through correspondence and the review of drafts. The signatories include both participants in the February 1986 meeting and other scholars who have reviewed and endorsed our conclusions and recommendations.

Our statement is not an exhaustive treatment of the issues but a summary of common ground. Our intent was to start with the most clear-cut findings on the critical issue of cost; to discuss those findings where partial evidence is available; and to suggest important and feasible research that would extend and enrich our understanding of the systemic price that America pays for the urban ghetto system.

II. OVERVIEW

Historically, black housing demand was channeled into either ghetto housing or housing in communities contiguous to the existing ghetto boundary. Blacks tended to pay premium prices for housing in neighborhoods adjacent to the ghetto. These neighborhoods in turn underwent rapid racial transition. Major dislocations were experienced by white households, local businesses, and community institutions. The housing stock passed through a cycle of overcrowding, declining maintenance, disinvestment, and displacement of the middle-class black newcomers by poorer residents. The result was an expanded urban racial ghetto—and the cycle began again.

The primary cause of this cycle was the strict separation of the black and white housing markets and the channeling of black demand into a very small sector of the housing market outside the ghetto. The malign consequences of the cycle were compounded when sales and rental agents, mortgage lenders, small businessmen, and government officials made no distinction between middle-class and poor blacks and thus helped to perpetuate a self-fulfilling prophecy of economic decline accompanying racial change.

Some of the costs of this historic system of discrimination are not difficult to recognize. The black middle class was charged a high price for its attempt to escape the ghetto because of the scarcity of appropriate housing options and denied the normal perquisities of middle-class status such as a home in a secure community where home values accrue and whose schools provide access to good jobs and further education.

Whites suffered as well. Whole communities and their local institutions dissolved. Long-established businesses, nonprofit organizations, and churches closed, sometimes forever. Financial institutions left communities, taking with them local investment capital.

The city as a whole suffered too, particularly when ghettos came to cover large parts of the municipal area. Middle-class taxpayers departed. The tax base shrank from the loss of business and employment in the ghetto areas and the deterioration and eventual abandonment of housing stock in later stages of the cycle. City agencies and police forces had to deal with the disruptions caused by the cycle as white communities mobilized to resist the beginning of racial change.

These costs of the classic ghetto system were overt and recognizable. Today, after a decade of experience with national and local fair housing laws, both the causes and the costs of housing discrimination and segregation are more difficult to detect and assess. Small-scale integration began in some previously all-white areas. Ghetto boundaries expanded rapidly, making more housing units available to black buyers and renters in many cities. As supply increased to meet demand, the relative cost of housing in black areas decreased, particularly in the oldest black areas, where demand was least intense.

A reverse black migration out of a number of northern states in the 1970s also affected demand pressure in the black housing market. During this period, the black middle class expanded; the black birthrate declined; and major policy changes, intended to redress the lack of capital for housing and business in black areas, were adopted. These changes, combined with the use of more subtle mechanisms of discrimination after the enactment of fair housing laws, rendered the consequences of segregation less visible and the mechanisms more obscure.

Our most basic conclusion is that none of the well-known economic, social, and statutory changes have fundamentally altered the ghetto system. It remains largely intact. It still restricts residence for millions of black families and thus limits the many benefits that residence brings. Ghetto expansion and resegregation of integrated neighborhoods are still taking place, although not always with the speed and predictability of earlier periods. In the core ghettos, isolation is more intense than it has ever been.

Moreover, certain consequences have become more dramatic as the urban industrial base collapses and employment opportunities migrate to the outer suburbs and beyond. At the heart of the ghetto system the older poor black communities now experience an unprecedented distance from white areas, from middle-class black areas, and from the rapidly growing sector of labor markets.

In short, the extreme and near-absolute segregation of the old ghetto system has given way to a more complex system whose dynamics are less apparent but whose consequences and costs are sometimes even more severe because of the general deterioration of the older central-city economies.

III. WHAT IS KNOWN ABOUT CONTEMPORARY HOUSING SEGREGATION

PERSISTENT SEGREGATION

Data from the 1980 Census confirm that most metropolitan areas remain highly segregated. Although there was modest improvement in the 1970s (following a decade in which metropolitan segregation actually increased), there has been almost no change in some of the largest centers of black population. The white population of large central cities is declining. Two-thirds of large urban school systems were predominantly nonwhite by 1980, and almost all were experiencing continuing racial change. Segregation was most intense in states with the largest metropolitan areas. Black students in New York and Illinois were more segregated than those in any other American state during the 1984-85 school year.

Research also reveals a substantial increase in the number of blacks living in suburbs—although their absolute number is still only a tiny percentage of suburban populations. But for most blacks, segregation remains severe—and the outward movement from the central city of whites portends increasing physical, social, educational, and economic isolation for ghetto residents.

PERSISTENT DISCRIMINATION

Discrimination is less overt but still pervasive in the nation's housing markets. A powerful new research tool—systematic audits of housing markets—has provided the best direct measure of current practice. The only national audit completed, by the U.S. Department of Housing and Urban Development, assessed market behavior in 40 cities. It found a high probability that a black family would encounter at least one case of readily measurable discrimination in the course of a typical housing search. (This 1979 HUD study did not address the problem of residential steering, which is the practice of showing blacks only homes in black or racially changing areas and directing whites away from integrated areas to all-white communities.) Subsequent audit studies and compliance investigations in a number of cities in the 1980s have shown a continuing high rate of discrimination.

SEGREGATION AND DISCRIMINATION FOR HISPANICS

Hispanics experience considerable residential segregation, particularly in large cities with large Hispanic populations. Audits of the Dallas and Denver housing markets found that Hispanics experience even more discrimination than blacks. Hispanic barrios are expanding rapidly in a number of metropolitan areas. The segregation rate, however, tends to be significantly below the black level of isolation. Young Hispanics are becoming increasingly concentrated in highly segregated schools. Hispanic segregation tends to decline with increases in income, education, and knowledge of English. Puerto Ricans are the most segregated of the Hispanic communities; their isolation in metropolitan New York City is at a very high level.

LENDING AND MORTGAGE FINANCE

Studies in several parts of the country have uncovered patterns of unequal mortgage finance practices in black, white and integrated neighborhoods. After controlling statistically for differences in income and housing cost, such studies have made it clear that white neighborhoods receive a disproportionate share of housing finance. Black and integrated areas are treated less favorably. One study shows that stably integrated areas are treated considerably better than racially changing communities. Several studies confirm claims by neighborhood groups that Federal Housing Administration (FHA) loans are now an important source of funding for mortgages in periods of racial transition.

FAIR HOUSING ENFORCEMENT

Federal, state, and local fair housing laws constituted a promise of significant change in access to housing. That promise has not been fulfilled. The research shows that discriminatory practices have persisted and that the laws have been weakly enforced. Projections made from housing audit research indicate that several million violations of fair housing law occur every year. The Justice Department, however, has never brought more than fifty cases for the entire country in any given year, and the number has been far lower in the 1980s. HUD still cannot enforce change on violators who will not voluntarily change their behavior. There is little or no enforcement of many state and local laws. Private action to enforce laws through testing and litigation is limited to a few metropolitan areas—where private civil suits in federal courts have proven highly effective.

If the housing industry voluntarily conformed to the law, the existing procedures for fair housing enforcement might be adequate. They are clearly insufficient to change a system in which discrimination remains entrenched.

PUBLIC ATTITUDES

A substantial majority of Americans support housing integration in principle. But most white Americans do not believe that discrimination is a problem and express little support for more assertive governmental action. Most blacks and whites have differing views of the proper level of integration. Whites tend to prefer integration with a large white majority, while blacks tend to see the ideal as a 50-50 neighborhood balance. Blacks express a very strong preference for integrated as opposed to all-black neighborhoods. Whites and blacks do not perceive a single housing market; they rather see communities as "color-coded" both by racial composition and acceptance of or hostility toward black homeseekers.

RACE AND SUBSIDIZED HOUSING

Policies designed to disperse subsidized housing had a substantial impact in some housing markets during the 1970s. Many units were constructed or rented in private buildings outside the ghettos and in suburbia. The potential impact on segregation, however, was not realized because units in white areas tended to be rented to whites. Even where site selection standards were enforced, there were no effective civil rights policies governing tenant selection. The overall impact of the subsidized housing constructed since the passage of fair housing laws is still to reinforce rather than remove residential separation.

IV. UNDERSTANDING THE COSTS OF SEGREGATION

When the facts of segregation are related to changes in other urban institutions, some of the costs of the system become apparent. In other cases, specific research is needed to unveil the linkages.

BARRIERS TO BLACK ECONOMIC PROGRESS

Because most black housing demand is confined to present and future ghettos and to rental rather than ownership markets, the wealth of both individual black households and the entire black community is diminished. Blacks are seriously underrepresented among homeowners, even when income is held constant; buyers in black neighborhoods have less access to capital for mortgages or home improvements; blacks who buy homes tend to buy relatively less expensive homes in areas where values increase less rapidly.

These patterns reduce the net worth of black families because equity in a home is the only substantial source of wealth for most American families. Reduced wealth in turn increases economic vulnerability to short-term reverses and shrinks such key opportunities as the ability to finance college educations and to trade up to more costly homes. A 1986 Census study showed that the typical white family has 11.5 times the wealth of the typical black family. The home ownership problem is one major reason why the gap in wealth is far greater than the gap in income.

A less obvious but equally significant cost is that the denial of home ownership in desirable neighborhoods forecloses access to better education and mainstream socialization for children. High school and college completion is a major factor in improving employment and income; where children go to school is a major factor in determining school achievement and completion. The differences among black, white, and integrated schools in terms of dropout rates and achievement scores are highly significant. Hence children of identical abilities confront highly differentiated opportunities that are a function of residence.

To the extent that different school and community environments decrease the likelihood of dropping out of school and increase the probability of entering college (both of which are also related to much lower probabilities of crime and welfare dependency), the value of housing opportunities plays a major role in determining lifetime earnings, taxable income for the community, and the demand on public purses.

ACCESS TO EMPLOYMENT

Since the 1960s, urban economies have been fundamentally transformed and work opportunities redistributed across the urban landscape. Industrial work has declined radically; most new industrial investment has been placed on the periphery of suburbia. For men without education, industrial jobs offered almost the only mobility and income adequate to support a family. Such jobs have declined by the thousands in many cities—and the new industrial jobs now being created are simply not accessible to workers in the inner-city ghettos and barrios.

A parallel and related phenomenon is the creation of huge tracts of abandoned or unmarketable industrial land within cities—land that is already fully provided with roads, sewers, utilities, and other expensive capital investments that must be made anew in the suburbs. Considering the location of new industrial jobs, the absence of public transportation links between the central city and suburban jobs, the location of the uneducated and unemployed minorities, and the school dropout rates of 50% among young central-city minority males, it is clear that residential segregation has an extremely high present cost that will multiply over time with an increase in the numbers of men unable to marry and support families.

The costs reckoned in terms of a city tax and service problems are compounded by the cost entailed in the depreciation of what should be prime inner-city land for economic development. If race is a basic cause of the disinvestment, this cost also would be related to residential segregation and the mechanisms of ghettoization. Industrial location studies do not typically ask about race as a variable, but it is clear that there have been very few new plant starts in black urban areas anywhere in the United States and that there have been massive closures of existing facilities. The location studies do usually point to such variables as safety and educational preparation of the work force and sometimes to the higher tax burdens that central-city facilities must pay for inferior services because of the heavy public safety and service burdens on central-city governments. All of these conditions are related to the syndrome of ghettoization.

The booming downtown sectors of cities that are successfully negotiating the transition to postindustrial economies provide very different kinds of jobs than those that traditionally have been the escalators of advancement for America's urban poor. There appears to be little spillover to nearby minority neighborhoods; many of the desirable jobs go to suburbanites whose schools retain and teach young people. Downtowns have therefore become a special metropolitan niche, offering low-wage service jobs to some minority workers and professional jobs primarily to suburbanites. Even in healthy cities, downtowns tend to account for a continually shrinking portion of the metropolitan labor market. It has become commonplace for healthy downtowns to exist near dying neighborhood economies.

Residential segregation is strongly linked to the highly restricted economic and social impact of downtown service sector growth. There would be obvious advantages to many of the two-worker small families that make up such a large proportion of the young white middle class and to the young unmarried workers to live close to downtown. The strength of the color line in the housing market excludes white demand from the affordable black and Hispanic neighborhoods near downtown—except in that small fraction of cases where gentrification produces a belief that the neighborhood will eventually be dominated by middle- and high-income whites.

An equally serious constraint on large-scale white use of such housing options is the lack of middle-class schools with substantial white enrollment; few whites with small children move into central-city neighborhoods with black or Hispanic schools. The aggregate cost of this syndrome for white workers is very high and is the cumulative sum of daily commuting costs, multiple car ownership, more expensive housing, and isolation from the rich variety of cultural and recreational opportunities found in a great urban center.

The very future of American cities is thus at risk. Middle-class settlement on a substantial scale would produce the capital investment needed to preserve and improve declining housing stock, commercial facilities, and neighborhoods. In an economy where more than nineteen in twenty housing units are in the private sector, and many of these units need continuing investment if they are not to be lost, there is no other realistic investment source. The tax revenues created by the combination of neighborhood reinvestment and family income would help to strengthen the city revenues in terms needed for the revitalizing spiral of urban change.

On a yet larger scale, the problems that affect the minority neighborhoods and create the irrational costs for those who would benefit from living in a convenient central-city location also affect the metropolitan area. Locations with easy access to the features of a city that led to its initial development, to the transportation and cultural centers of the region, and to the community's own historic roots are simply not considered for new investment. Most young white families never even consider living in the city. A vast duplicate infrastructure is created outside the city at a very high cost and new housing units are constructed even when the number of households stops growing.

While many forces in addition to race drive suburbanization, race is clearly one of the most powerful. America's color line, embodied and embedded in the dual housing market, not only creates barriers for blacks and Hispanics; it also creates barriers for whites so invisibly effective that they are not even perceived as barriers. Neighborhoods and schools and other institutions in areas that are all nonwhite are taken completely off the list of choices for the white and black middle class. As neighborhoods continue to change, the boundaries of this unknown and excluded area come to embrace more and more of the entire city. This racial dimension is a basic reason why city neighborhoods cannot effectively adapt in many cases to a changing economy and social structure.

The system of ghetto creation and expansion is clearly related to the deterioration of older black neighborhoods. Once fair housing permits ghetto boundaries to expand more rapidly, the demand for housing at the core of the ghetto must depend on the continually replenishing immigration into the city. As the black middle class moved steadily outward and both black fertility and migration from the South changed dramatically, demand at the core of the ghetto dropped rapidly, and the degree of concentration of the very poor increased. With the core ghetto closed off to white housing demand and experiencing disinvestment of many kinds, economic and social deterioration accelerated. What the white outward movement was doing to the central cities as a whole, the black middle-class outward movement and the lack of replacement residents were doing to the older black communities.

Many of these trends cannot now be measured with any precision. But the nature of the social and economic dynamics involved in the processes of urban segregation is clear and the existing empirical research on specific issues and specific locations is compatible with these questions. There are, however, a number of questions that deserve more complete research.

V. RESEARCH NEEDS

While most of the issues and problems addressed in this statement could benefit from additional research in a variety of settings, some among them have received relatively less systematic analysis than others and yet occupy a central role in advancing our understanding of contemporary ghettoization. These include the following:

(1) *Residential Steering*. Steering is the practice of showing minority and white homeseekers with the same qualifications and interests housing in different neighborhoods. It is probably the most serious form of contemporary discrimination. It is rarely detectable by individual homeseekers, and individuals have no way of pooling their experiences to document the existence of steering. Steering needs careful study in a variety of housing markets.

(2) *Impact of Enforcement Strategies*. The question of whether or not fair

housing enforcement policies change housing market practices is of central importance but has not been studied extensively. We need to learn more about the relative efficacy of litigation and other strategies of enforcement.

(3) *Discrimination and Segregation Involving Hispanics.* Hispanics may become the nation's largest minority population within a generation or two. We need better information on Hispanic housing preferences, sources of housing information, and the nature and severity of discrimination against Hispanics in housing markets.

(4) *Housing Market Knowledge Among Blacks and Whites and the Effect of Counseling.* A crucial legacy of overt, de jure segregation is that blacks and whites have very different kinds of information about the housing market and both view housing situations in light of past experience. Given that limits of personal knowledge and the practices of sales and rental agents produce housing choices that reinforce segregation, we need to understand the potential of techniques designed to broaden the range of housing choice and to encourage "nontraditional" moves.

Do people want such information? What is the most effective way to provide it? Should potential clients be shown unfamiliar areas and housing by a fair housing advocate? Is there any problem of discrimination against blacks or whites who wish to move within the traditional boundaries of racial change? Knowledge about these issues would show if counseling can help maintain integration and reduce the costs of the ghettoization process.

(5) *Psychological Costs.* Housing discrimination and ghettoization affect an individual's psychological well-being. What happens when a black family conforming in every way to middle-class standards learns that it has been excluded on purely racial grounds or that it faces harassment and vandalism after moving? How can one calculate the costs to older white residents of friendships broken up, local institutions dissolved, and cherished neighborhoods abandoned with high financial and personal costs?

(6) *Self-Fulfilling Prophecies of Disinvestment.* To what extent do businessmen, lenders, rental property owners, and other investors believe that racial change produces economic deterioration? How are these expectations related to actual investment decisions and how often does a self-fulfilling prophecy develop?

(7) *Community Life Cycle After Racial Change.* What is the typical sequence of economic and educational change following racial change in city and suburban communities that have become all black or all Hispanic in recent years? What variables account for differences in the pattern? What proportion of middle-class minority homeseekers end up in stably middle-class minority communities and under what conditions are such communities most likely to stabilize?

(8) *Racial Change and Local Employment.* Neighborhoods passing through racial change frequently experience loss of some businesses and the downgrading of the others. We need broader and more specific studies of the nature of the

changes, the reasons for the owners' decisions, and the consequences for local employment. Such studies might well also attempt to quantify the impact on the cost of living when less expensive chain stores are replaced by lower-volume, higher-cost ghetto merchants. The businesses studied should include professional offices such as medical providers.

(9) *Changes in Home Ownership Levels*. Systematic comparisons of levels of home ownership in single-family residential communities before, during, and after racial transition could provide more specific documentation of the relationship between ghetto expansion and mortgage finance discrimination. The shifts from owner-occupancy to rental management and changes in housing-based family wealth should also be tracked and assessed.

In addition to these nine specific areas where more systematic research is needed, we need briefly to define three broader research needs:

(1) DEVELOPMENT OF BETTER MEASURES AND BASIC DATA

Researchers would greatly benefit from the development of certain basic data for all major housing markets. The calculation of black-white and Hispanic-white segregation measures on block and tract levels for all U.S. housing markets should be completed. Several different measures, including the dissimilarity index and the exposure index should be computed; so too should frequency distributions, showing the distribution of the total black, white, and Hispanic populations among Census tracts of various levels of minority concentration.

In order to relate racial composition and racial change to more detailed housing and community-level data and to permit greater understanding of changes occurring between Censuses, more use should be made of the Annual (American) Housing Surveys sponsored by HUD and the Census Bureau. At present, these surveys have limited value for fair housing research because they do not record the racial composition of the tracts from which survey responses are collected. That information should be added to the data so that many additional relationships, particularly those relating to neighborhood-level issues, can be studied.

In addition to the standard measures of the randomness of population distribution, basic data should be calculated for the last two Censuses showing the change in physical distance between typical young white and minority families in urban housing markets. Plans should be made for rapid computation of all these measures shortly after the release of the 1990 Census to avoid the long delays in the provision of crucial policy information following the last three Censuses. The dramatic decline in computation costs should make it feasible to provide better information much faster than in the past.

(2) SYSTEMATIC STUDY OF INTEGRATED COMMUNITIES

The great force and dramatic effects of the ghettoization cycle explain why so much research has addressed this process and why so little has been directed to the study of stably integrated communities. Much of the research literature has in fact ignored the subject. Integration has been defined as nothing more than a transitional condition.

That definition reflects the dominant historical experience. But the fair housing movement and the development of fair housing law seek the development of a different housing market, one with free transactions in all directions, where the race of newcomers does not threaten communities, and where integration will be commonplace. A number of communities have been stably integrated for some time. If research is to help realize the goals of federal law, we must understand the conditions that produce and sustain integrated communities and their effects on minority and white families and institutions.

A crucial step toward understanding the costs of segregation and the value of its alternative, integration, is careful comparison of integrated neighborhoods with other neighborhoods that passed through a variety of racial transitions. The following issues should be compared over the short and long run: educational levels, local school test scores, occupational status of residents, home ownership, housing code violations, rehab investments, numbers and kinds of neighborhood jobs, continuity or change in major community institutions, community capacity to attract middle-class home purchasers with children, racial attitudes of adults and children, and public beliefs about neighborhood quality and future.

Data of this sort, collected over a considerable period of time, would add significantly to our knowledge of the costs of the ghetto system and the effects of achieving lasting integration.

(3) THE IMPACTS OF GOVERNMENTAL POLICIES

Most housing segregation research focuses on individual choice and ignores the consequences of governmental action or inaction. Key research tools such as public opinion surveys, audits, and econometric models of market behavior tend to treat choices as unconstrained decisions, except where clear evidence of discrimination is found. The structuring of choice by public policy has been largely overlooked.

Governmental policies that can shape racial outcomes in housing markets involve the location and tenanting of subsidized housing and the counseling of subsidized and other homeseekers; zoning and planning; mortgage lending; fair housing and real estate certification; school desegregation; maintenance of property; and services in integrated areas.

There are a great many racial data collected on a number of these issues, including subsidized housing and the racial composition of schools. Research could show what fraction of low-income minority families live where they do because government decided to provide or approve the provision of subsidized housing in one place rather than another. It could test the hypothesis, supported by some research, that far-reaching school desegregation plans, covering much of the housing market, facilitate stable residential integration in neighborhoods. By analyzing both overall patterns and important exceptions, such as the court-ordered effort to allow Chicago low-income families to live in subsidized units in private suburban apartments, research could assess the degree to which segregation results from public rather than private choices and could illuminate the circumstances under which public agencies can achieve more favorable impacts.

VI. ISSUES FOR THE FUTURE

Some important research questions require such comprehensive information on urban development that they cannot now be analyzed with any precision. Is it true, for example, that the direction of black migration within a metropolitan area affects the pattern of economic development on a metroplitan scale, directing funds away from one sector and to the periphery of other sectors? Or is it true that predominantly minority cities decline faster than those with less racial change? So many factors shape such outcomes, all of which reinforce economic decline in the central cities and in some outlying areas, that it is difficult to assess accurately the specific impact of race and of racial discrimination.

There are many such questions—all of great complexity—to be answered if we are to make continuing progress in sorting out the wide-ranging impacts of discrimination in urban housing markets. Much work remains to be done by the nation's research community. All of the participants in the Chicago meeting agreed that it should be done—because urban housing discrimination and segregation are central structural features of both race relations and contemporary urban development. If we are to disassemble these structures in the interests of a healthier and more harmonious urban future, we need both to know more and to act decisively on what we do know.

About the Contributors

THOMAS A. CLARK is Associate Vice Chancellor for Academic Affairs and a member of the faculty in the College of Design and Planning at the University of Colorado at Denver. He previously served on the faculties of McGill, Middlebury, and Rutgers. He is the author of *Blacks in Suburbs: A National Perspective*, a forthcoming book entitled *Community Economic Development*, and articles on regional development, housing policy, neighborhood process, and structural change in urban labor markets.

JOE T. DARDEN is Dean of Urban Affairs Programs and Professor of Geography and Urban Affairs at Michigan State University. He has done extensive research on the housing problems of racial minority groups in metropolitan areas. His works include *Afro-Americans in Pittsburgh: The Residential Segregation of People*, and several articles in scholarly journals. His most recent book is *The Ghetto*, published by Kennikat Press.

JOHN E. FARLEY is Professor of Sociology and Research Associate in Regional Research and Development Services at Southern Illinois University at Edwardsville. He is the author of two books, *Majority-Minority Relations* (1982) and *American Social Problems* (forthcoming, 1987). He has also published a number of journal articles, has given expert testimony in the federal courts concerning housing segregation and discrimination, and has served as a consultant to Central Mortgage and Housing Corporation (Canada).

JAMES W. FOSSETT is Assistant Professor of Political Science in the Institute of Government and Public Affairs at the University of Illinois. A former staff member of the Brookings Institution, he has published widely on urban policy matters and is currently investigating the problems of the urban poor in obtaining access to primary medical care. His most recent book is *The Changing Politics of Federal Grants* (Brookings, 1984).

SUSAN GOERING is one of several attorneys who represented school children from the predominantly black Kansas City, Missouri School District in their school desegregation suit against several governmental entities including the U.S. Department of Housing and Urban Development (HUD) and the State of Missouri (*Jenkins v. Missouri*, No. 77-0420-CV-W-4, U.S. District Court for the Western District of Missouri, filed May 27, 1977). Goering is currently Legal Director of the Maryland chapter of the American Civil Liberties Union (ACLU).

JULIA L. HANSEN received her Ph.D. in economics from the University of California at Berkeley in 1984. She is currently an Assistant Professor of Economics at the University of Colorado at Denver. She is the author of "Housing Problems of Asian Americans" in J. Momeni, ed., *Race, Ethnicity and Minority Housing in the United States.*

DAVID J. HARTMANN is Assistant Director of the Community and Family Study Center and is a Ph.D. candidate in the Department of Sociology at the University of Chicago. He is writing his dissertation in the field of urban sociology. He has written about managed integration in the Chicago suburb of Park Forest South.

FRANKLIN J. JAMES is an Associate Professor in the Graduate School of Public Affairs of the University of Colorado at Denver. He directed HUD's Legislative and Urban Policy Staff during the Carter administration, and has been a senior member of the research staff of the Urban Institute and the Rutgers University Center for Urban Policy Research. He received his Ph.D. in economics from Columbia University, and is the author of numerous books and articles on urban and housing policies.

JOHN F. KAIN is Professor of Economics at Harvard University, where he has been a member of the faculty since 1964. During 1975-1981 he served as Chairman of Harvard's Department of City and Regional Planning. Professor Kain received his A.B. degree from Bowling Green

State University in 1957 and his M.A. and Ph.D. degrees from the University of California at Berkeley in 1961. He is the author of numerous books and articles on urban transportation, housing markets, housing administration, and neighborhood change.

BETH J. LIEF is an attorney who served as Assistant Counsel with the NAACP Legal Defense and Education Fund, Inc., for eight-and-a-half years, where she litigated civil rights cases involving housing discrimination, school desegregation, employment discrimination, and access to health care facilities. She later was Executive Director of the New York City's Mayor's Special Education Commission and worked at the New York City Board of Education in a policy-making position. She is currently a Program Director at the Edna McConnell Clark Foundation, in charge of projects regarding the homeless, minorities, and New York City.

GARY ORFIELD is Professor of Political Science, Public Policy, and Education at the University of Chicago. He has served as a consultant to numerous federal, state, and local agencies and prepared studies for HUD, several state legislative and congressional committees, the National Institute of Education, the Education Commission of the States, and other agencies. He has been an expert witness or court-appointed expert in numerous school and housing desegregation cases. He has written widely in the areas of school and housing segregation, Congress, national social policy, job training, college access, and other issues. He received his Ph.D. in political science from the University of Chicago.

YALE RABIN is Professor of Planning and Associate Dean for Academic Affairs in the School of Architecture at the University of Virginia. His research, over the past twenty years, has been concerned primarily with the impacts of the land use-related policies and practices of government on low-income and minority groups. His findings have influenced the outcomes of precedent-setting civil rights litigation dealing with school segregation, housing discrimination, exclusionary zoning, and discrimination in the provision of municipal facilities. He has served as a consultant to the U.S. Department of Housing and Urban Development, the U.S. Commission on Civil Rights, agencies of state and local government, and national public interest organizations concerned with equal opportunity.

GARY A. TOBIN is the Director of the Cohen Center for Modern Jewish Studies at Brandeis University in Waltham, Massachusetts. He earned his Ph.D. in City and Regional Planning from the University of California at Berkeley. Dr. Tobin has published extensively in the areas of housing, population research, and social planning. He is the editor of *The Changing Structure of the City* and *Social Planning and Human Service Delivery in the Voluntary Sector.*

LOUIE ALBERT WOOLBRIGHT is a statistician at the Office of International Statistics in the National Center for Health Statistics. His major interests are social change, demography, human ecology, modernization and development, and urban sociology. His dissertation at the University of Chicago was about Hispanic, black, and Asian residential succession and transition. He has written on population change and evolving patterns of segregation in the Chicago metropolitan area.

JOHN YINGER is a Professor of Economics and Public Administration at the Maxwell School, Syracuse University. He has also taught at the Kennedy School of Government at Harvard University and the Institute of Public Policy Studies at the University of Michigan. In 1978-1979 he served as a Senior Staff Economist at the Council of Economic Advisors. Most of Professor Yinger's research concerns cities. He has published theoretical and empirical articles on urban housing markets, often with a focus on racial discrimination, on the fiscal condition of cities, and on the interaction between the housing market and the market for local public services.

NOTES